GRAVITY

CONVERSATIONS WITH G

A COMMON DIALOG ON
UNIVERSAL GRAVITATION AS
" THE SECRET "
LAW OF ATTRACTION

A GRAPHIC NOVEL OF PURE
ABSTRACT (NON) FICTION

★ FEATURING THE @Joms™

VOLUME II

GRAVITY: CONVERSATIONS WITH G
A COMMON DIALOG ON UNIVERSAL GRAVITATION
AS "THE SECRET" LAW OF ATTRACTION

FEATURING: A GRAPHIC NOVEL OF PURE ABSTRACT (NON) FICTION

→ THE @Joms™

BY: Joie

★ UNIVERSAL GRAVITATION.org

★ GRAVITOLOGY.org

PRODUCTIONS:

```
10101010   10    10    10101010 ®
10     10  10    10    10     10
10     10  10    10    10     10
10     10  10    10    10     10
10     10  10    10    10     10
10101010   10    10    10101010
```

www.OLIOSTUDIOS.com

SCIENCE = GRAVITY

ISBN: 1468019279

LCCN: 2011962215

CONTENTS

2 - IN G WE TRUST

3 - ACTS OF G

4 - POWERS INVESTED IN G

5 - RATED G

PUBLIC DOMAIN:

- Blavatsky, Helena : The Secret Doctrine

- Darwin, Charles : Origin of Species

- Kepler, Johannes : Harmonices Mundi

- Newton, Isaac : PRINCIPIA + OPTICKS

- Ovid (43BC-17AD) : METAMORPHOSES

O

UNIVERSAL GRAVITATION

AS

"THE SECRET"

LAW OF ATTRACTION

J01E

"If you pardon me, I shall rejoice ☺!
The die is cast, and I am "WRITING"
the book — to be read now — or by
prosperity; it matters NOT.

As G itself has waited over 6,000
years, and now FINALLY has
witnesses."

— Johannes Kepler
Transcribed from:
"Harmonices Mundi"

"The most beautiful system... may only proceed from the counsel and dominion of a powerful **INTELLIGENCE**."
—Isaac Newton (1687)

Why do some think INTELLIGENCE may only describe Atomic Structures—like ourselves? Would not magnificence resemble the ineffible **SPACE** where we are contained? Daniel Monti, in "Down the Rabbit Hole," discussed our universe as being a LIVING SYSTEM in which we are the sums of its parts.

4% ← Here we are, condensed within ≈ 96% INVISIBILITY

THE UNIFIED FIELD

We know the least about **SPACE**, yet continue seemingly trivial pursuit of its contents (the visible 4%), while ignoring the fundamental **dark** order of our **HOST**.

What does it mean to be made of celestial molecules within a terrestrial world? Why are Atoms mostly empty **SPACE** —hollow enough to marinate maximum **CHI** or lifeforce?

Earth also shares our human attributes at a grand scale of fluids + stuff circulating within a thin membrane that consumes, digests, and transforms fellow Atoms. Chemically speaking, people are the Earth — its Atoms, colors, and flavors. If Gravity or "G" brought the planet together, and continually holds "US" as the part(icle)s within, how else may it be affecting or EFFECTING species?

To say **G** is solely responsible for **falling** towards centers is to not consider its complete

powers of motion that includes rising as with heat or evaporation; pushing or separating as repulsion (e.g. divorces, supernovae); pulling or sucking (e.g. gravity wells, black holes), and **attraction** in general as binary stars, people, and other Atoms are drawn together in terrestrial relationships.

In addition to harnessing power for modern life, we know about **ELECTRICITY** in relation to **US**, and have examined its many forms — from electrons- to photons; lightning in the sky; and our very own **BIOLOGICAL** wiring that sparks inside of us and eelfish relatives. Plus, **MAGNETISM** has been examined thru James Clark Maxwell's discovery and Michael Faraday's initial research.

Yet, **Gravity** has remained hidden for centuries — only showing partial views of itself in direct relation to the observer's quest. Dean Radin's "STUPIDITY HYPOTHESIS" may also play a role in limiting further exploration of the **dark** force (page 113) that continues to illude us.

This body of work, termed **GRAVITOLOGY** (201) explores Universal Gravitation in the fields of astronomy, chemistry, biology, meteorology, geology, psychology, (meta)physics, other sciences; and most importantly, **US**. Before now, humans have never linked themselves as the Atoms of Attraction according to Carl Jung's **SYNCHRONICITY** or Steven Strogatz's, "SYNC." How can we assume **G** BONDS the planet + stars, but not **US** as the water people living within a **SPACE TANK** (69)?

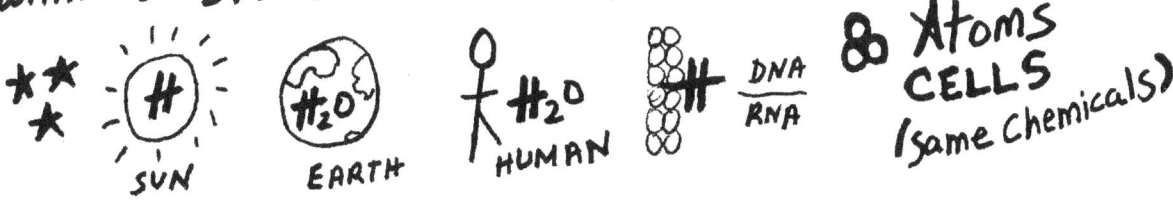

G has been given numerous names, although character **EFFECTS** are the same as Casimir, Coriolis, London, Great Attractor, Strange Attractor (chaos Theory), Covalent or Ionic Bonds, **DARK** (Matter, Energy, Flow...), and so on. Maybe the **CONSCIOUS** Quantum Computer activity described by Stuart Hameroff is also Gravity disguised as the so-called Van der Waals force?

In Volume 3, which combines PRINCIPIA, OPTICKS, and THE SECRET DOCTRINE, Newton references:

"The force which retains celestial bodies in their orbit has been hitherto called CENTRIPETAL, but it's now being made plain and clear that it be **NO** other than the **GRAVITATING** force with **CENTRIFUGAL** power that we shall hereafter call **GRAVITY**.

Gravity endures forever; is EVERYWHERE present; and by existing always, it constitutes duration and **SPACE**; a **TIMELESS** infinity where Atoms are contained therein...

Atomic attraction and impulses are the same...a "calling" means where the answer results in **MOTION** as bodies tend toward another."

G is **SPACE**; and it also exists thru numerous faces and places as **ENERGY**, which is naturally gravitational : GE = MC² (167).

As Gravitational Energy (GE), Gravity may be the Active Principle that Sir Isaac duly searched for in ALCHEMY, an early chemistry. **GE** is the only **POWER** with supermassive "Strong" and seemingly "Weak" characteristics to generate action or Butterfly Effects;

it is the sole Fundamental Force where Atoms are employed to **WORK** via Laws of Motion in the Quantum Field where "**SPOOKY**" action may occur at any distance.

ENERGY is normally defined as an ability to **WORK** - usually over a distance - with movement. However, as the MOTION MASTER, equivalent to acceleration, **G**ravity fits the definition of ENERGY itself, even outside of its 'potential' or 'kinetic' power; it seems to be the **NULLPUNKTENERGIE** sought by Max Planck and Albert Einstein; and the vacuum **AETHER**, which those like Nikola Tesla attempted to harness.

The silent **SPACE** of Zero POINT GE may also be the same Ø's located on the "Critical Line" of Bernhard's "Riemann Hypothesis" that refers to Prime Number distribution. Note that Prime Numbers are like Atoms where all other digits are composite molecules. The Riemann Zeta function correlates with **G**ravitas as the Casimir Effect incognito; and as **ENERGY** levels of Atoms in Random Matrix Theory where the ZEROS provide attractive or repulsive **SPACE.**

$$\sim (\exists r : \exists s \ (P\ (r,s) \land (s : g\ (sub\ (f_2\ (y)\ )\ )\ )\ )\ )) \quad KURT\ GÖDEL$$

The Uncertainty Principle of Incompleteness Theorum is also a part of life; yet, by exploring **UNIVERSAL GRAVITATION** as the FORCE that brings Atomic Structures together in EVERY form, "The Secret" Law of Attraction (79) may be unified with other sciences.

As David Hilbert would say, "Wir müssen wissen; wir werden wissen." Otherwise, how can we ever achieve a 'Theory of Everything' if humans are excluded?

We are the Atoms!

@Joms™

ACT **1**

Conversations With

Joie

ORIGIN
OF
SPECIES
& &

"Who can explain the essence of Gravitational Attraction?

...I see no good reason why the views given in this volume should shock the religious feelings of anyone.

It is satisfactory, as showing how transient such impressions are, to remember the greatest discovery ever made by man — namely the Law of Attraction of **Gravity**...

There is a grandeur in this view of life with its several powers, having been originally breathed by the Creator into few forms or one; and that, whilst this planet has gone on cycling according to the fixed Law of Gravity, from so simple a beginning, **endless forms** most beautiful and wonderful have been, and are being **evolved**."

— Charles Darwin

I

Holy G

The Atoms
Gravitationism
Creationism
Ovid
Colors

@Joms™
· The Atoms ·

★ We pledge allegiance to the Atoms, and the united states of which we individually stand... One **SPACE**, indivisible, with *UNITY*, and **G**ravitational **E**nergy for **ALL**. ★

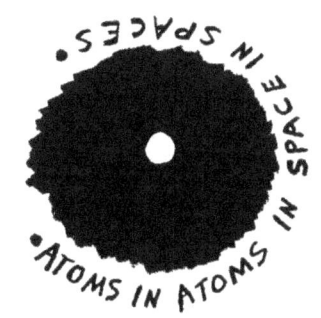

Aloha...

Bula!

We are the Atoms, **HRH** His + Her Royal Hydrogen

The word "Atom" (our Name) is used for simplification, although Atomic Structures come in many shapes, colors, sizes, and terms—from the tiniest 'NEW' sub**atom**ic particle—to the grandest stellar or celestial machine.

Atoms descend from a noble heritage of gases, liquids, and solids. In Human, Flora, Fauna, Microbial or celestial form, we're usually saturated with some sort of Hydro or "**H**"lineage. Hence the saying, "For every Atom, there exists an equal or opposite."

 Was it Godfried Wilhelm Leibnitz ~~+~~
who invented the Wunderbar Binary System
(0,1) on behalf of G ?

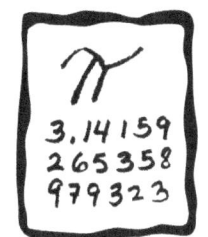

We're neither (math)maticians
...nor numerologists,
but we [LIKE] numbers, letters, + symbols.

There's a great deal to (dis)cover, so let's
ENTER the next scene.

·Gravitationism·

$G \rightleftarrows 0$

"In the beginning was the word."

The word was **G**;
The number **Z**ero.

Then, **G** said,
"Let there be LIGHT!"

Holy **G**!
This is when **H** replied, "I'll **G** first."

Forever typecasted, **H**YDROGEN has been and may always be ☆superstar☆ of virtually every show.

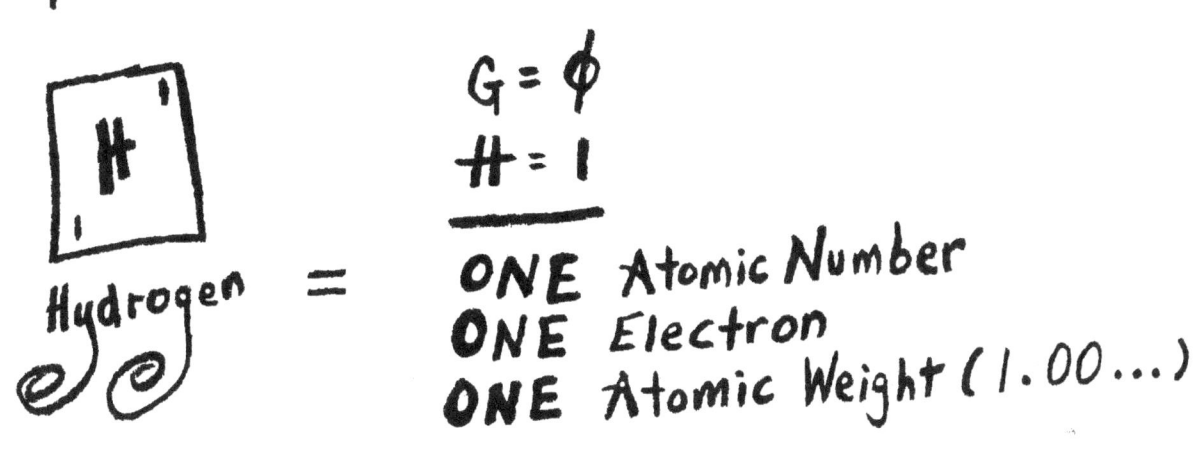

$$G = \phi$$
$$H = 1$$

Hydrogen =
ONE Atomic Number
ONE Electron
ONE Atomic Weight (1.00...)

H is our great ancestor...
the first Λtom in which we're all
CONNECTED as "one" cosmic family.

Ever since **H**, we are no longer
created, per se, nor destroyed;
We just trans (form):

H = Adam | **He = Eve**
(HYDROGEN) (HELIUM)

H reveals a great deal about
us; We drink and breathe it as a
daily requirement. Plus, **H** grows
our food, bonds our **DNA**, and
so much **more** (53)!

96%　　　　　　　　　　4%

First = $\emptyset$ zero$\underline{\text{th}}$ order → GE SPACE

Next = I Hydro → Helium →
and Other Atomic
　　　　Relatives

In the beginning was the word.
The word was G.
Then, came H of which all Atoms are "one."

@oms™
•CREATIONISM•

Why do you think Atoms create(d) this planet?

We are **Creationists.**
All Atoms are Creators...

Our existance alone "creates" Earth, and everything we do, say or think is a form of

CREATIONISM

that Gravitates
unto itself.

@Joms™
•OVID•

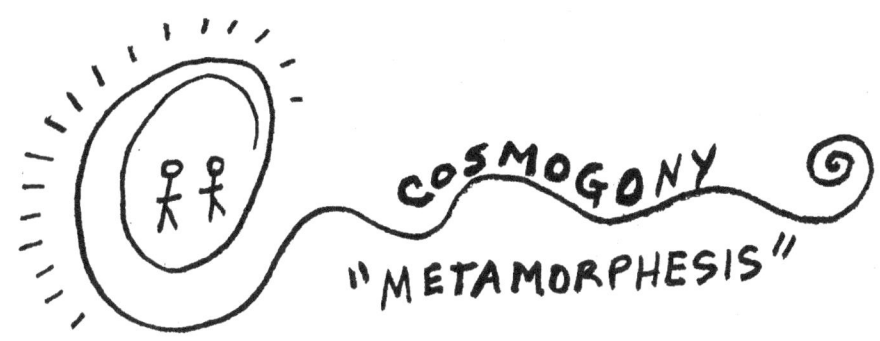

COSMOGONY

"METAMORPHESIS"

THE CREATION

"Before the ocean, Earth, or heaven,
Nature was all ALIKE; Shapeless
Nothing but **DISCORDANT ATOMS**...

U? ♀ ⋮ ♀ No, Us!

G settled all argument and seperated...
So, things evolved and found each other,..
BOUND in Eternal [Gravitational] Order,"

Is this Newton's recipe for "The Net"
Or the alchemical cosmic web?

@Joms™
• COLORS •

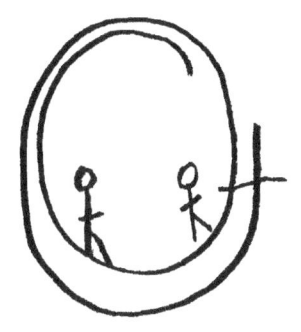

Atoms are not only Creationists, We're **Artists** too!

How so?

Every move we make with #— including the Seagulls and their Butterfly Effects— (re)paints the globe.

EARTH

We are the **COLORs** in gravitational MOTION.

→← ↑↓ ←→ ↓↑

2

In G We Trust

Gravitational Cycle
Qué Ondo?
SPACE TANK
100%
Warticles

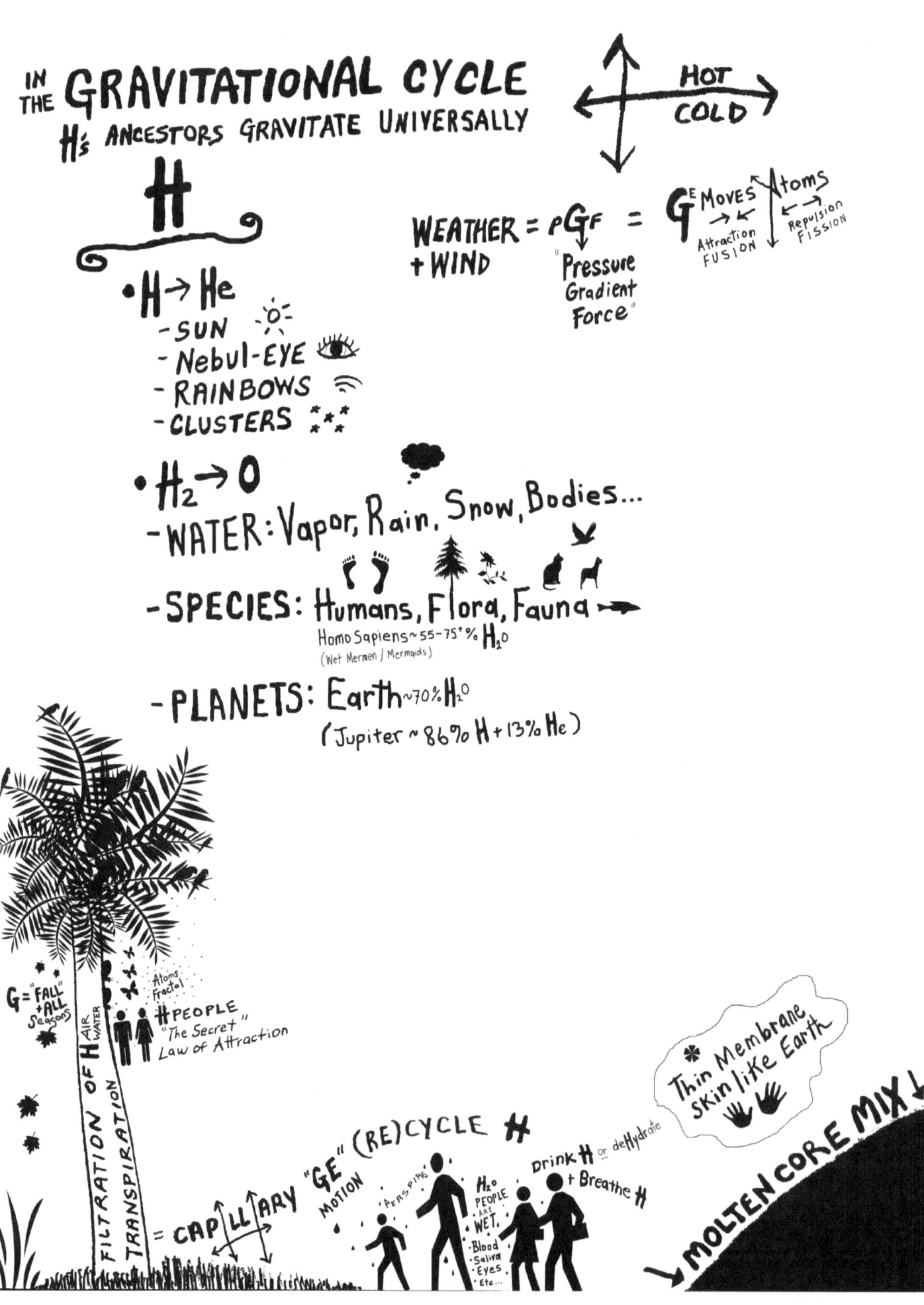

ATOMIC BIFURCATION

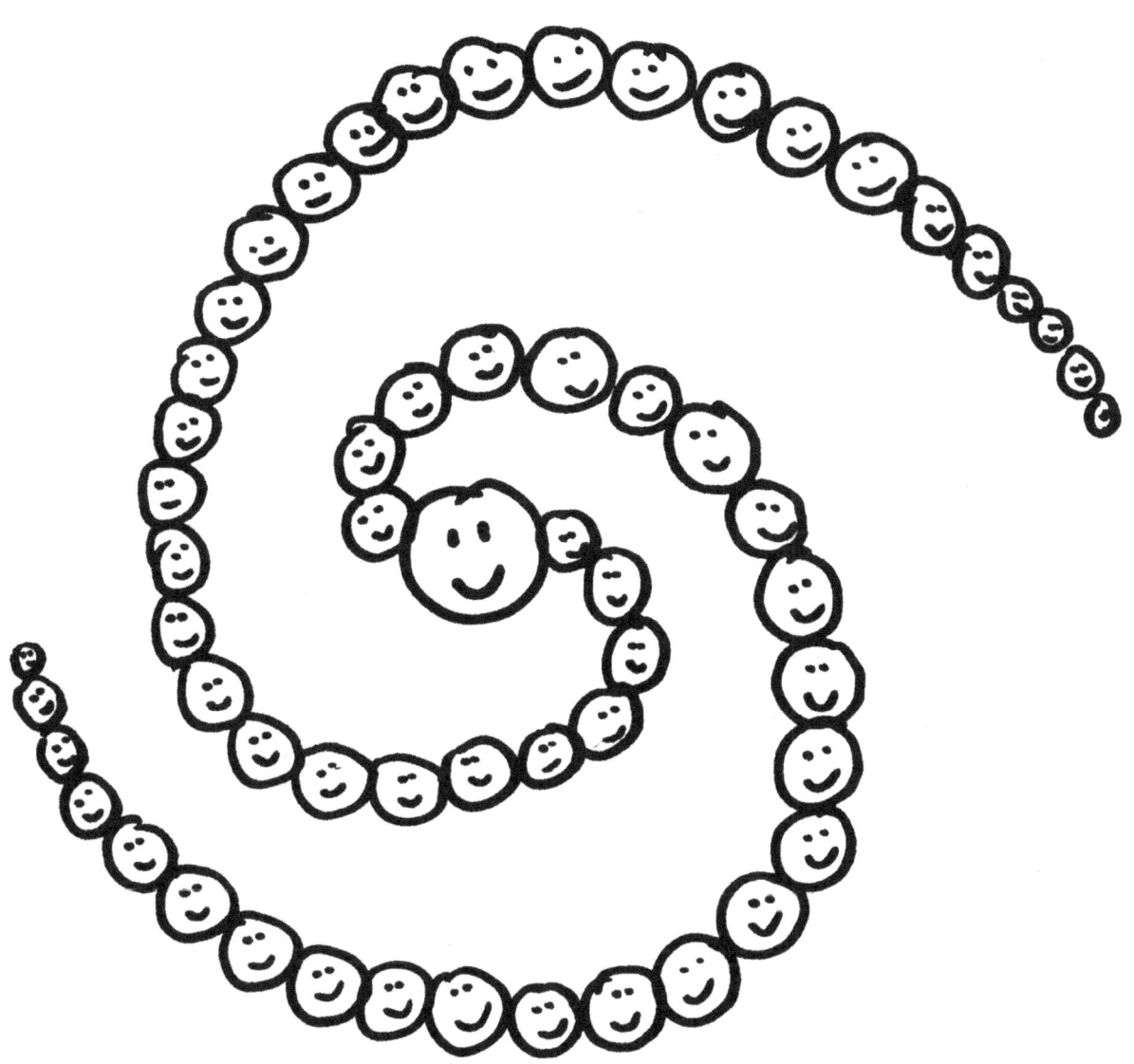

GRAVITATORS
GRAVITATE
GRAVITATIONS

@OMS™
• GCYCLE •

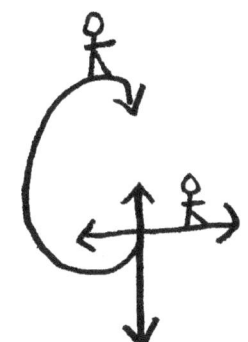

As ENERGY, **Gravity** is the dark shapeshifter with Laws of Motion that operate the **Gravitational Cycle**, and controls the **H**ydrologic.

Since we're all related to **H**ydrogen, Atoms naturally interact and gravitate across SPOOKY distances in **SPACE**(s). In fact, gravitation may be thermal, nuclear, electrical, chemical, and more considering the entanglement of Atomic relationships.

In the SPACETANK (69), Di**H**ydrogen Oxide may be a gas, liquid, or solid wherein Atomic Structures (53) are fueled machines that aid in **H**'s circulation.

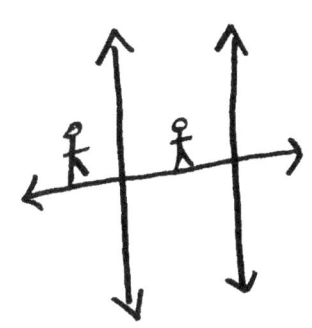

In fact, Hydrogen is Gravity's #1 employee, filling bodies of all shapes, sizes, + scales; it works as GE's agent to weather, sculpt, and flow in perpetual motion with other famous relatives.

Strangely, **H** appears to gaze at itself from afar thru the lens of its Nebul-EYE.

*** In super clusters or BI-TRI-QUAR-NARY
** systems, **H** dances in gravitational circles.

● **H** forms Blackholes that are found in the CENTER of most every galaxy (163).

Thru **Hydrogian Evolution** (predating Darwinism), **H** expanded into planetary formation, where it still dominates matter of all shapes and sizes.

Most can't survive without it, as the superstar also grows our food with **Gravitational Energy (GE) Power.**

In addition to BEAMING GE with Vitamin D, H is well-known for growing flora as PHOTO(N)SYNTHESIS. Stellar warmth is loved to such a degree that large numbers of Atomic Structures flock to areas on Earth with the most sunshine. Plus, SUNTANS are popular too. H also shines to gravitate its H_2O TIDES via the MOON'S reflection (179); and it radiates brightly on SUNNY NIGHTS.

TRIVIA:

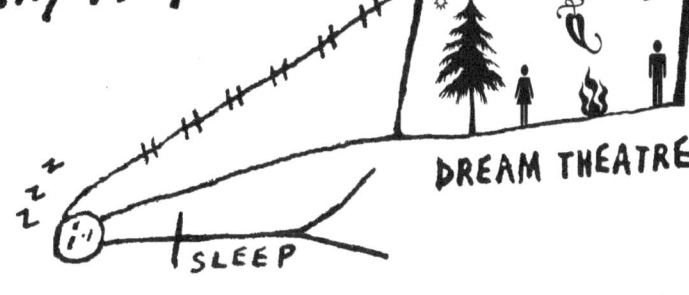

DREAM THEATRE

SLEEP

★ Do Humans in slumber project MOTION PICTURES using realtime H-PHOTONS for internal lighting on the holographic mind screen?

HYDROGEN AVAILABLE NOW!

RATED #1 SINCE CREATION.

ALL PURPOSE USE FOR GE POWERED MACHINES.

FUELS:

* * SUNS + STELLAR SYSTEMS

* O • PLANETS + CELESTIAL BODIES
  (COMETS + METEORS TOO)

* • FLORA + AGRICULTURE ♻

* • FAUNA + PETS 🐾🐕

* • VEHICLES + ZEPPELINS ✈

* • HOMO SAPIENS = CONSUMPTION REQUIRED OR DEHYDRATION MAY OCCUR

WETTEST FISH (Mermaid + Mermen) IN THE H_2O SPACETANK (69):

PEOPLE ARE THE GASEOUS PART OF THE H_2O SPACETANK

* WAXES EAR DRUMS + HEARS
* SALIVATES TO SPEAK, LICK, TASTE, SWALLOW, OR TONGUE KISS
* LUBRICATES FOR INTERCOURSE
* BLOOD FOR CAPILLARY ACTION, CIRCULATION, OR MENSTRATION
* URINATES, AND MUCH MORE

THE GASEOUS PART OF:

* BLINKS TEARS + SEES
* BREATHES WATER VAPOR
* FARTS H_2O [TAILPIPE EXHAUST]
* DRIPS MUCUS OR PUS
* PERSPIRES, CRIES, + SPITS

HUMANS LOOK LIKE ⚤ in solid, liquid, + gas for motion ---

SAILABLE • SWIMMABLE • DRINKABLE

BLESSABLE : HOLY WATER

STATION

FREE

$H_1, H_2, H_3 ...$

H_2O has "Hidden Messages" says, Masaro Emoto.

H + Oxygen (O), Carbon (C), Nitrogen (N)...

GE groups — as we isolate ourselves - or mix and match. Differences are generally separated, not solely by **MASS**, but via "ENERGY~ FREQUENCY that stems from thought with Humans.

Countries (Map Lines or Boundaries), Cultures, Species, and More are divided like **RAINBOW** colors of electrical contrasts.

HOT ↑ Even Heat + Cool temperatures are gravitationally ——— DIVIDED by resonance. Warm, fast degrees COLD ↓ eventually **MEETUP** with cold density to produce

WIND, according to the P "G" F, which is Gravity again described as a Pressure Gradient Force in Meteorology- Plus, the "Great Ocean Conveyor is self-similar.

Alone with H_2, "O" magically appears in us all as solid, liquid, or invisible **vapor** where it may gravitate upward to form clouds, only later to descend as **fog**.

For most of us, H_2+O is a daily requirement for survival and capilliary action, transpiration, perspiration, evaporation, and much more. In fact, inside our moist brain, we see thru water with wet eyes.

G CYCLE

H_2O → gravitationally "FALLS" as rain, snow, tears, and urine; it washes **air** of dust, soot, or other particles; and is the same drinking water that composes most bodies in **solid** form.

Duly employed as Glaciers to store itself frozen, **GE** streams H downward to our homes, rivers, lakes, and oceans; it's nutritiously filtered thru gravel, sand, and mud before seeping into springs or aquifers.

Glacial Ice Sheets are also hired to SCULPT or **design** mountains, valleys, boulders, rocks, and land in outsourced partnership with Plate techtonic shift and subduction.

To assist with continental desire, within Earth's molten core, land ingredients perpetually mix before gravitating up/out to baked perfection.

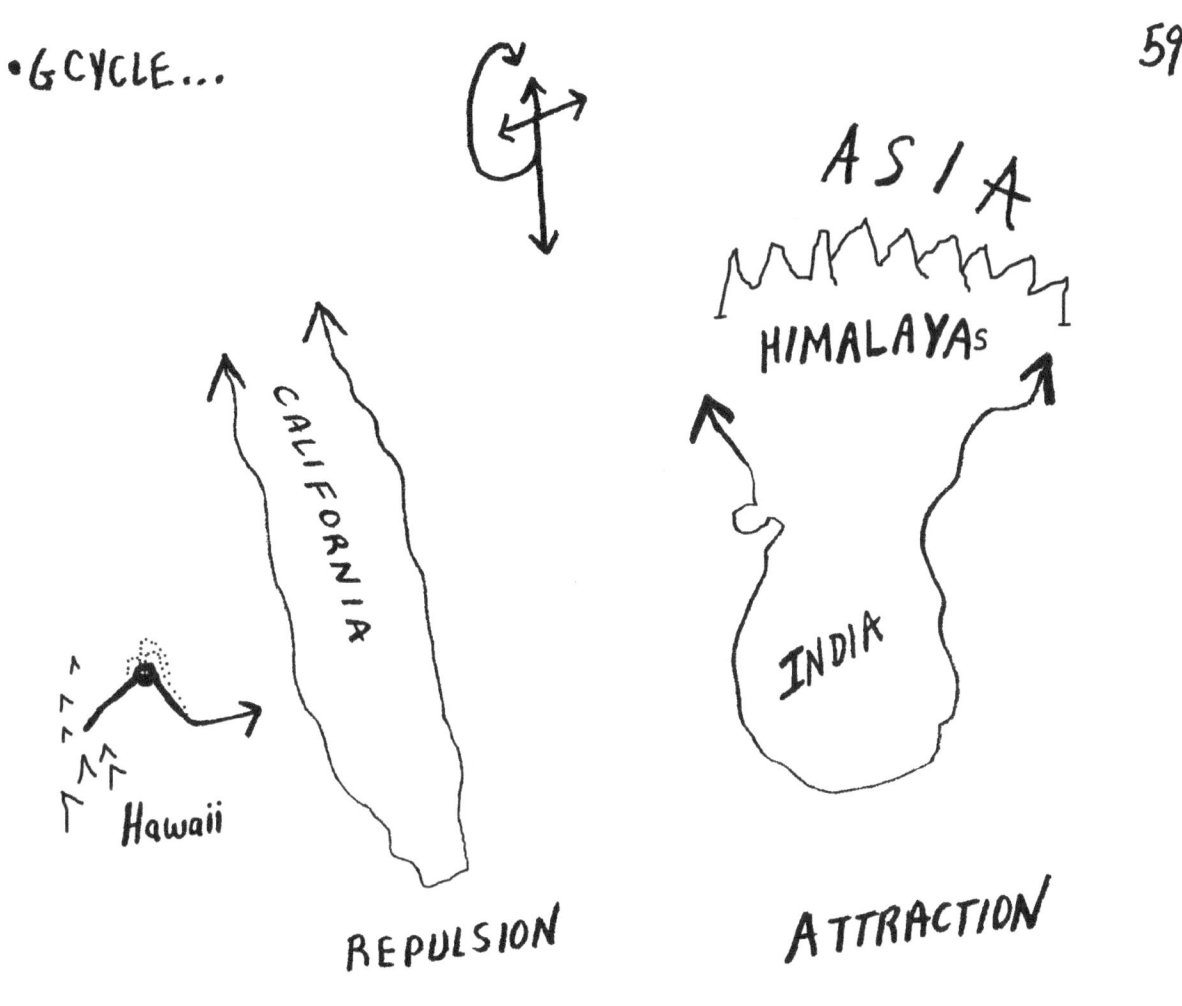

While India's "attraction" to Asia brought countries together, the desired island of California "repulses" Northward, as Hawaii joins the mainland United States with all other Atoms forever in (trans)formation.

TRIVIA:

Is **Gravity's** equivalence to "Acceleration" the natural effect of RAPID **EXPANSION** [especially since **SPACE** remains **constant** throughout Atomic evolution]?

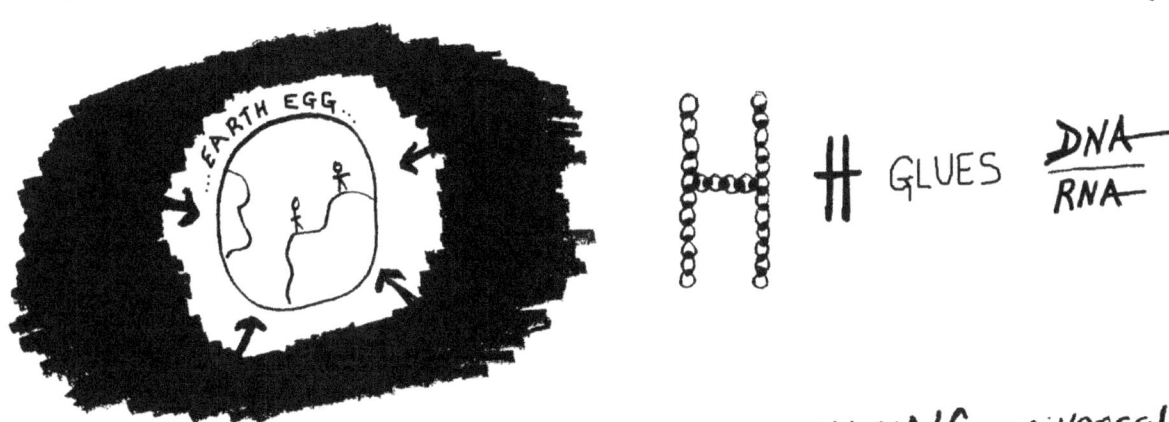

Here we are, GRAVITATED and GRAVITATING universally within our planet SPACE TANK (69) with H_2O, He,N,C,Ca,N,Na, S,Fe,Ne,Mg,Si... + relatives fueling GE powered machines.

In addition to being PHOTOgenic, H people are PHOTON-lit, G'ing each other through "The Secret" Law of Attraction (79). Whether we breathe H on land or under sea, our bodies are mostly Hydro with "empty" Gravitational Energy SPACE.

Our babies grow in H water, just like fowl or fish EGGS, which Earth duly resembles in the dark womb of GE SPACE. Plus, relative to our planet's atmosphere and EGGSHELLS, our body is fractally emerged in water contained in a thin skin membrane.

In Optics, Newton reminds us that "Eggs grow from insensible magnitudes, and change into animals; tadpoles into frogs; and worms into flies.

All birds, beasts, fishes, insects, trees, and other vegetables grow out of Water... and by putrefaction return again into watery substances."

→ ← $\updownarrow$ ← →   ↑ ↓ ↑ ↓ $\leftrightarrow$

Interestingly, as chemicals attract or repel based on their **ELECTRON** configuration, people also seem to gravitate via "LIKE" vibrations. In fact, our closest relations — even if equal or opposite — are usually **FRACTALS** of ourselves.

E-MOTION in the Gravitational Cycle attracts **REAL** physical molecules, as Candace Pert shows. So, $GE = MC^2$ by feelings alone! Therefore, are we simply gravitating ourselves in others who produce self-similar (phermone) chemicals?

In 'GENERAL SCHOLIUM' of PRINCIPIA, Sir Isaac discusses God, which we define as the infinite invisible **SPACE** of Gravity operating the GCYCLE; printed as "**G**":

"G governs **ALL** things... eternal, infinite, absolutely PERFECT; the same **G**, always and **EVERYWHERE**, in which **ALL** are contained and moved. Yet, neither affects the other; G suffers **NOTHING** from the motion of bodies whom find **NO** resistance from its spacious omnipresence Whence also G is all similar: all eye, ear, brain, arm and **POWER** to perceive, understand, or act in a manner **NOT** at all **H**uman, nor corporeal; but **UNKNOWN** to us.

As a blind man has no idea of colors, so we have **NO** idea of the manner by which the **ALL-WISE G** perceives and understands all ---

G is utterly VOID of all body and bodily figure; and can therefore neither be seen, heard, touched; **nor** ought to be worshiped."

G fits the description of **SPACE** itself where Albert Einstein postulated Atoms to warp or curve (121).

Gravitational Energy exists invisibly all around containing us as we CYCLE through it, as part(icle)s of it; whatever it is... it's evidently everywhere. Can you see **SPACE** in + outside of Atomic Structures (71)?

$$Z \rightleftharpoons Z^2 + C$$

As in Julia, Koch, or Benoît Mandelbrot's set, if GE equates to Z as Zero ($Z \rightleftharpoons \emptyset$), then Atoms are the colorful $Z^2 + C$ on the other side of this equation, fractally evolving with $\leftrightarrow$ Butterfly Effects to infinity.

Technically written, $GE \rightleftharpoons MC^2$ represents the IN / OUTPUT ITERATION loop of US Atoms with Zero Point Gravitational Energy.

¡Ola! ★ Qué Ondo?

How's your Gravitational Energy Waving?

Me   Buena!

Know, I wonder... if we didn't have sooo much in **COMMON**, would our relationship (193) exist?

Gravity's Law of Attraction seems to match or **GROUP** self-similarities (fractals) within the vast **SPACE TANK** of diversities, and Atoms often default our own relationships in the same manner (191). **Why?**

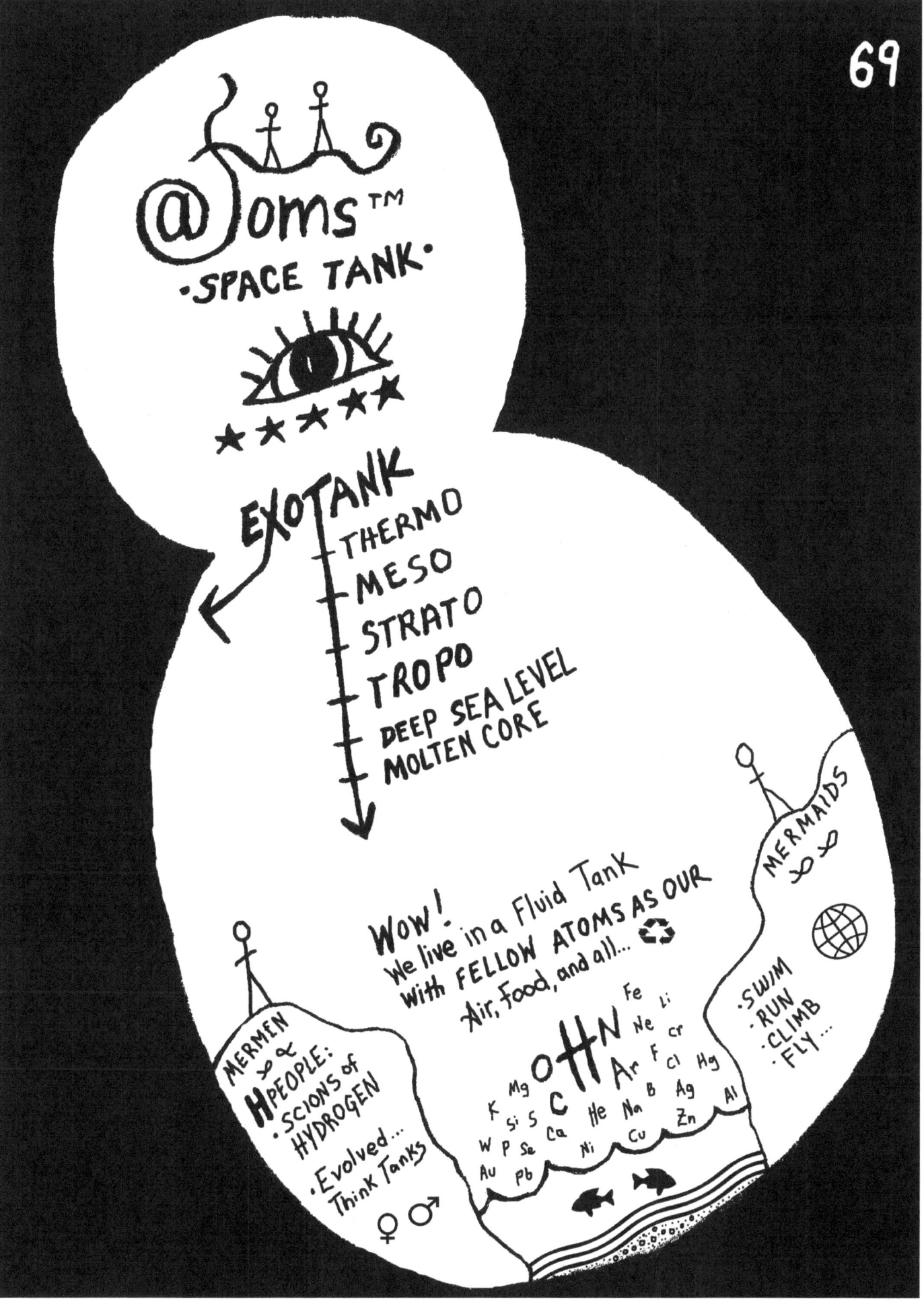

@Joms™
•100%•

U know how everyone insists that (Atoms⁺) the Universe is mostly empty?

Yep, it's evident everywhere we l👀K — even on this page!

The (white) **SPACE** is between every grass blade, water droplet, and all else.

We are 100% *WIRELESS*; inflated with invisible Gravitational Energy (97). **GE** is everywhere. Can U see it now?

Since Atoms are 99% 'Holy Spirit' **SPACE**, our SPIRITUAL life is evident. However, in **GE** we need not trust; gravitation is universally democratic; and all lifeforms(91) attract as members of a super-extended Hydrogen family.

@oms™
•WARTICLES•

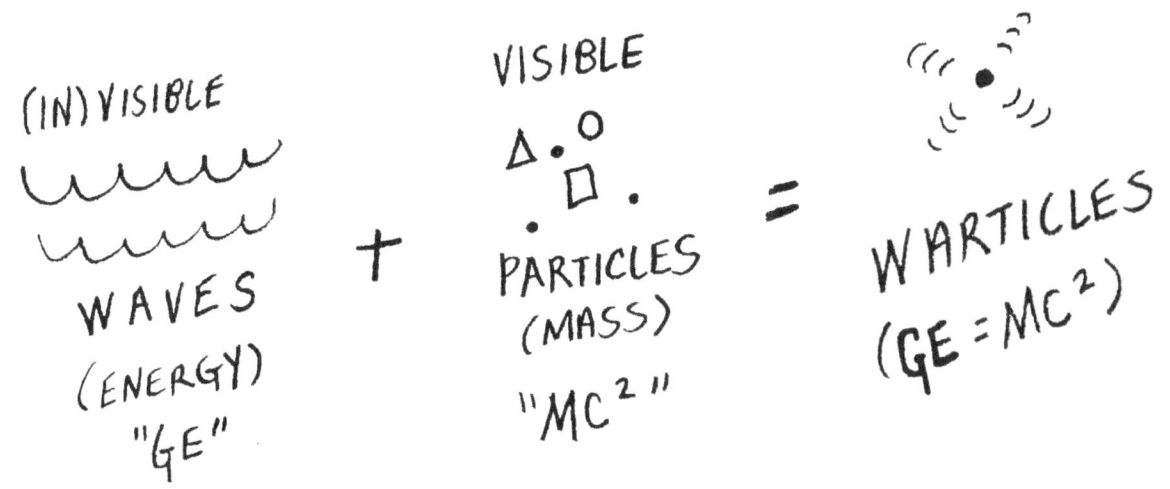

(IN)VISIBLE

WAVES
(ENERGY)
"GE"

+

VISIBLE

△ · O
· □ ·
·

PARTICLES
(MASS)

"MC²"

=

WARTICLES
(GE = MC²)

Atoms are both Physical (mass) and invisible Gravitational Energy, which is attractive in nature by its mere existance.

In "We Are All **WEIRD**", Seth Godin told the story about a Zoo that used **DATA** to "draw" crowds into its establishment.

Drawing, pulling, or attracting is Gravity's signature; it brings Atoms together in its **SPACE** on every Scale.

Whether the medium is physical or invisible words and **INFORMATION**, Atomic Structures affect (121) each other with potential & Kinetic **GE** EFFECTS.

3

Acts of G

The Secret
Desire
Dead or Alive
G SPOT

(PRO)VERB: As Atoms THINKETH or BEETH,
So shall Gravitation Attract or Repeleth.

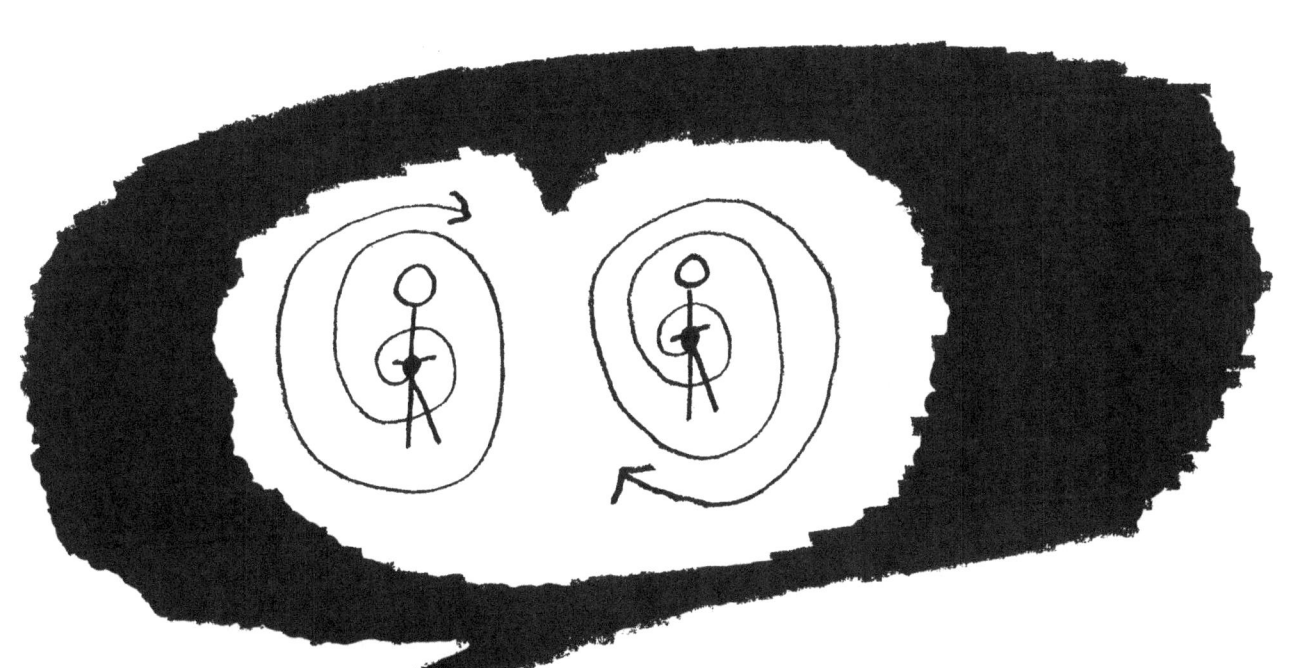

Atoms say, "The Secret" Law of Attraction (LOA)
is UNIVERSAL GRAVITATION; it's
gravity's private SPACE that's not visible
to the general public; it's EVERYWHERE
containing and bringing Atomic Structures
together around the cosmic web.

SEE: Volume **I** (212)

People are interested in LOA for evolutionary reasons; that includes **CHANGE** and manifestation. In essence, knowing the Secret enables one to gravitate their true desires by using focus of mind, which shifts attractions.

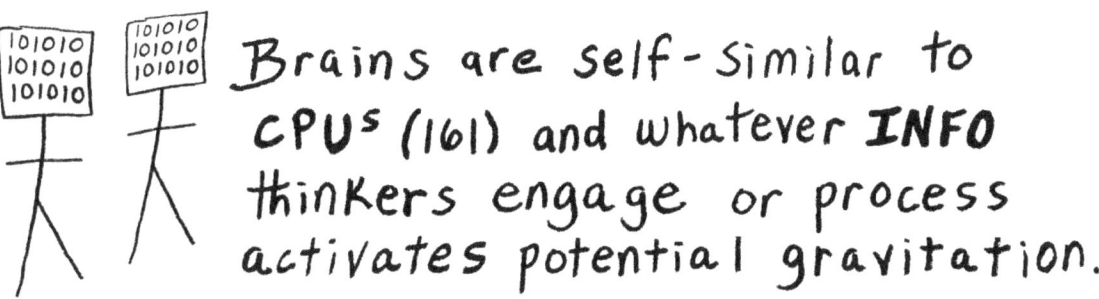 Brains are self-similar to CPUs (161) and whatever **INFO** thinkers engage or process activates potential gravitation.

Joie is known as the first visionary to link "The Secret" Law of Attraction with the science of **GRAVITY** ("GE Theory"). VISIT:

* The Secret. tv

 * Abraham-Hicks.com

* Universal Gravitation.org
  GRAVITOLOGY.org

Umm!
That was gooood.

...Let's get some MORE!

We should STAY HERE.

There must be a better place out there with **more** of us...
   Mo Betta **BUGS** & FOOD we never tasted!

....Yeah, we're supposed to be FRUIT FULL and multiply. ♡♡

Ya'll can stay here talkin.
I'm gonna go hoppin.

I gotta feeling we should stay.
RRRibbit!

CU later Alligatas!

RRRibbit!
Are U visual-EYE-zing again?

I have a dream that one day...

...What's that sound?

• DESIRE...

SSSHH !!

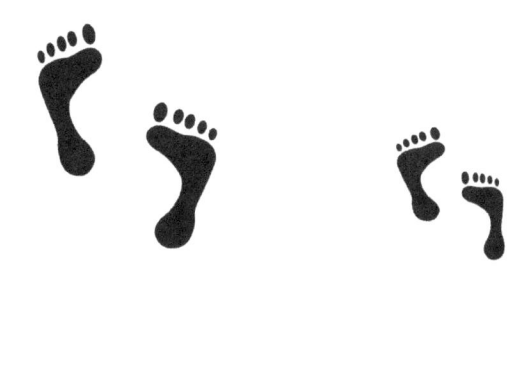

SSSHH !!

WHAT HAPPENED?

WHERE ARE WE?

2 WEEKS
2 DAYS
Later

Hear that ?.... *RRR*
RRRIBBIT
RIBBIT ...
RRR

PLAY TOAD MANSION
Gordon Vale, AU

𝄞 RRRIBBIT ♪ RIBBIT ♪
𝄢 RRR ♪   R R R   R R R
#

I KNEW IT!

.....AMPLEXUS?

BASED
☀N A
TRUE STORY
(OF DESIRE)

Wow!
DESIRE = Atomic Evolution
Even Bufo Marinus Expansion!

@oms™

• DEAD or ALIVE? •

Can inanimate substances create "LIVE" animate beings, like US?

Good question!
They say the first Atom was HYDROGEN ("H") that fuses →← to form HELIUM (He); and other elements.

If all composites gravitated from an "inanimate" Origin of Species, like H, then who's to say which one of us is NOW more ALIVE than the other?

Like the Asian Carp's DESIRE for flight, or the story of BUFO MARINUS (83), we all may be scions of Hydrogen that includes Oxygen with its wet and dry properties, which works in unison with others like stable Carbon.

Humans with our supercomputer brains may be highly evolved Atoms of chemistry involved with Creationism (35) of even MORE lifeforms.

(IN)ANIMATE = ANIMATE

@Joms™
• GSPOT •

What do you think about the popular
G SPOT?

You must be referring to **G**ravitational **M**ystery SPOTS...

What's so MYSTERIOUS about such enigmas?

After all, we live on a floating ball that's orbiting in **SPACE.**

Anything is possible!

4

Powers Invested in G

ZPGES

(IN)VENTION

CZASOPRZESETRZEN

SPACE TRAVEL

OFF THE CHART

@oms™
· ZPGES ·

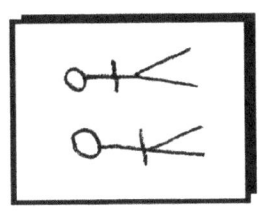

You know... Sleep, meditation, and transition take us to the same **SPACE** of *NULLPUNKTENERGIE*; it's the **Zero Point Gravitational Energy** that *Atoms* breathe in + out (within *Air*); it's where we originate, return, and plug-in to re-charge:

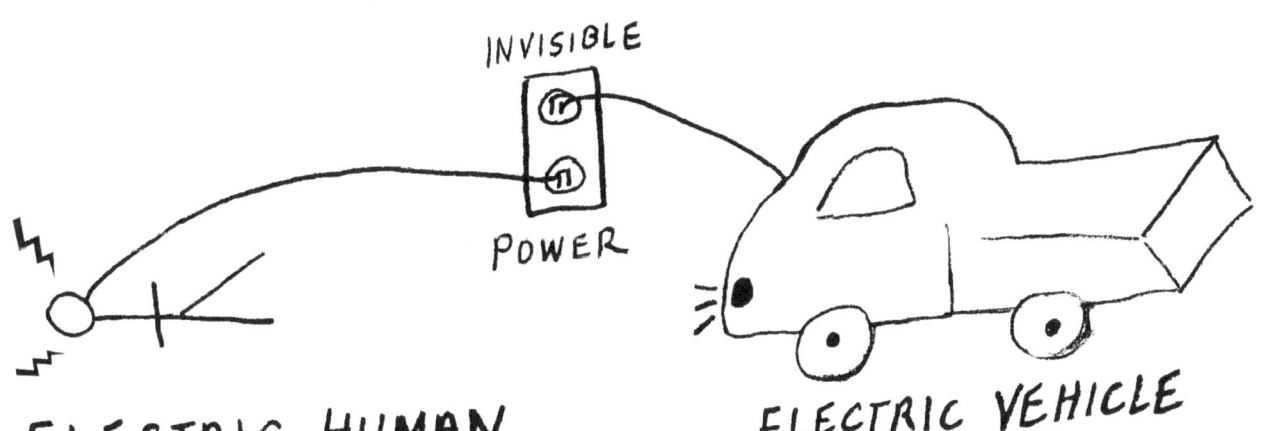

INVISIBLE

POWER

ELECTRIC HUMAN
WIRELESS RE-CHARGE
(e.g. Sleep, Meditation)

ELECTRIC VEHICLE

Wow! Same Source Energy
(**ZPGE-SPACE POWER**)

@oms™
• (IN) VENTION •

INVENTION #1: EEM
ELECTROPHORUS ELECTRICUS MOBILE

−600+ VOLTS

⚡ELECTRIC EEL MOBEEL⚡

Introducing the new EEL... the first vehicle of its kind powered by EELS inside the H_2O FUEL CELL TANK.

@oms™

.CZASOPRZESETRZEN.

DAY DARK

Night is a natural state of the cosmos without Atoms; it's invisible SPACE.

DAYLIGHT
H→He

SPACE is the dark Gravitational constant, in contrast to Atomic Structures, whom are always in flux within its omnipresence.

Even the term, CZASOPRZESETRZEN (SPACETIME) is misleading, as the word EARTHTIME would be more suitable to represent a specific place and time within **our** planet.

Celestial and Terrestrial bodies have unique gravitas, Relativity speaking. Thus, Atomic Time should be referenced like DOG YEARS, Relative to the Atoms within specific Gravitational Fields.

Examples: Earthtime, Moontime, Marstime, and so on.

SPACETIME itself represents darkblack infinity where we all reside in different forms.

In, "What On Earth is Wrong With G?" the host showed a TIME 'CONTROL' center, and it's interesting how SPACE itself is timeless; it's our container.

In fact, it's not really 2012, as was revealed in, "Do You Know What Time It Is?" All we know for sure is this moment may be either =

DAY LIGHT O   OR   ● DAY DARK

As with all INVENTED calendars around the world, Monday thru Friday and Weekends are falsified facts - only useful for organizing data, schedules, and MEETUPS - Relative to observers.

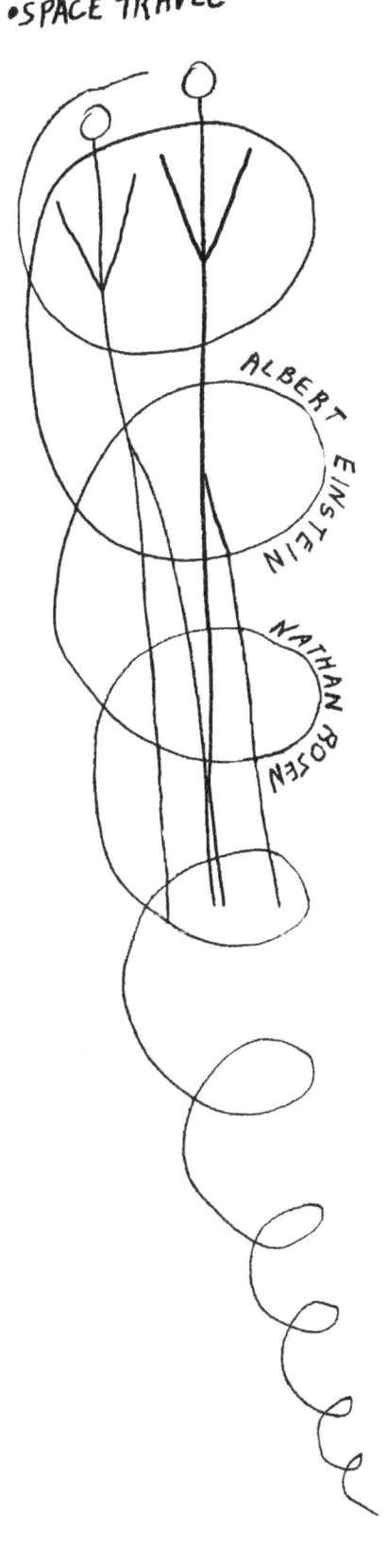

@Joms™
•SPACE TRAVEL•

ALBERT EINSTEIN

NATHAN ROSEN

The hypothetical Wormholes referenced by General Relativity takes us all over space and time, usually thru unseen (proxy) tunnels.

In fact, we can be **HERE** and **THERE** within **LIGHTSPEED**; yet electronic or wireless connections are unstable, especially thru the World Wide Web.

EVIDENCE:

- Internet communication is traversible, allowing 'REALTIME' video chat across continents.

- Robotic contact + control @ great distances is possible, such as thru Earth—Mars operations.

- Carl Sagan's "**CONTACT**" made no reference to current technologies that allow humans to connect online across the Einstein-Rosen Bridge:

$$ds^2 = \left(c^2 - \frac{2GM}{r}\right)^{-1} dt^2 + \left(1 - \frac{2GM}{c^2 r}\right)^{-1} dr^2 + r^2(d\theta^2 + \sin^2\theta\, d\phi^2)$$

Cosmicholes are evident, although we engage them (non)physically through various devices and invisible **SPACE** travel.

@oms™

·OFF THE CHART·

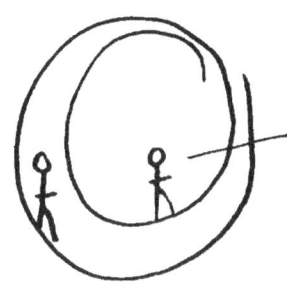

Did U know the PERIODIC TABLE was once CIRCULAR prior to the standard Rectangle FORM?

"Nature's Building Blocks" says there have been SEVERAL HUNDRED types throughout history.

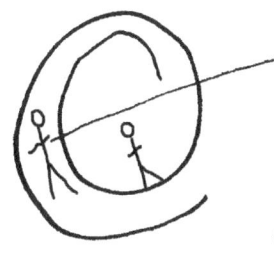

My favorite is Walter Russell's "SPIRAL CHART OF THE ELEMENTS."

EARTH

Since the beginning of EARTH TIME, all Atoms that began within our Home sphere haven't been destroyed or created... We just keep CHANGING FORM.

New Information = New forms

5

RATED G

ATOM GAMES
DARKSPEED
G-SPIN
DECISIONS^2
AD INFINITUM
Warp. Curve. Bend

@Joms™
•Atom Games.

FOR ATOMS
BY ATOMS

At the Atom Games, electrons/**photons** flash **on** as a **lit** image, and **off** as invisible SPACE. So, the questions for players are:

• As scions of Hydrogen, are **photo**graphic memories lit by **photons** thru the **photo**electric effect of internal **photo**receptors?

• Would Atoms be able to see, hear, smell, taste, or touch anything **NOT** made of Atoms?
(HINT: SPACE)

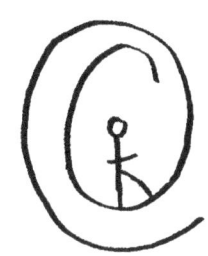

What's the opposite of "visible" LIGHTSPEED?

"INVISIBLE"
DARKSPEED

As the constant, ever present, timeless **SPACE**, there's **NO** need to travel when you're EVERYWHERE. Spooky actions of the Unified field may be performed:

• INSTANTLY — as the body of all.

• RELATIVELY — to variable speeds of Atomic Structures, such as 'strong' Black Holes **or** super Quantum force at the Planck Scale, in contrast to seemingly 'weak' **GE** with other matter(s).

@oms™
•GSPIN•

Does our latrine spin like the **TOP** or **BOTTOM** Hemisphere **?**

?? We'll see ??

Interestingly, our brain has two hemispheres submerged in H_2O also. Plus, like all Atoms, Earth and us have electrical: on/off QBITS controlled by **GE** (not Van der Waals), with microtubials, cells, + neurons flowing information **in/out:**

○ OFF ↓↑ ! ON

@Joms™
• DECISIONS²

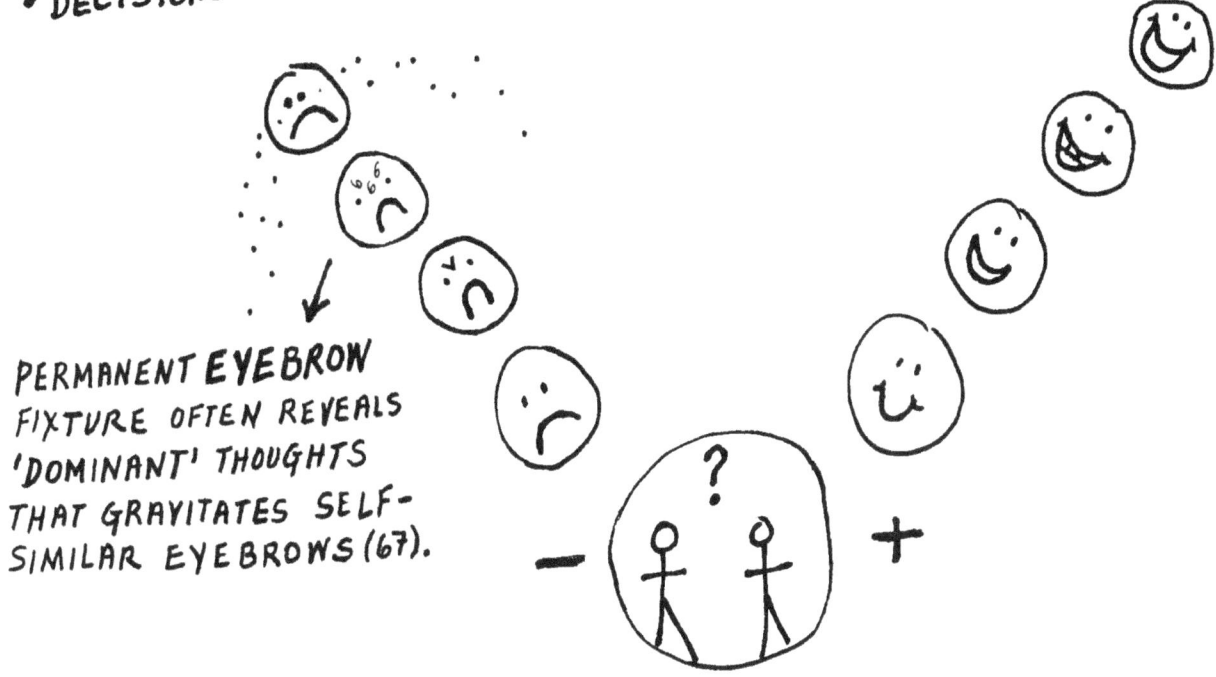

PERMANENT **EYEBROW**
FIXTURE OFTEN REVEALS
'DOMINANT' THOUGHTS
THAT GRAVITATES SELF-
SIMILAR EYEBROWS (67).

In, "*Think Like A Champion*," Donald Trump discussed the GRAVITATIONAL PULL of challenges, and how personal computation of **INFORMATION** ultimately determines the solution or outcome.

Since we're always processing data, our decisions FRACTAL, and thoughts of the thinker's choice EXPAND with potential or kinetic manifestation in Gravitational SPACE.

(e.g. ☺ = Happiness > Joy > Bliss...)

♪ ☺ "When You're Smiling"
LOUIS ARMSTRONG

@Joms™

Georg Cantor ☾ ✶

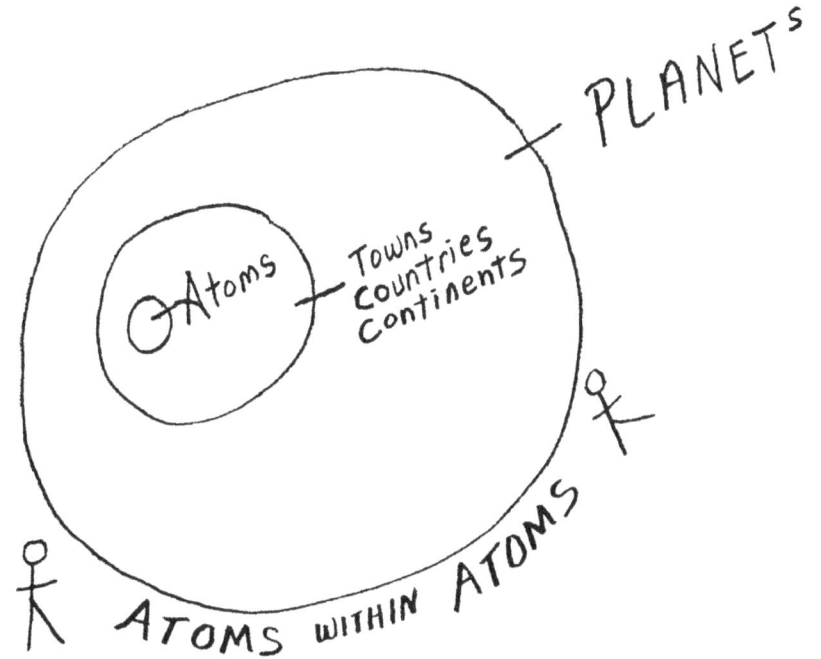

ATOMS WITHIN ATOMS

∞ **SPACE** as Gravitas infiltrates *ALL*; it's everywhere within everything and everyone as the 99% ingredient of Atoms; it does not age or transform as ordinary matter; it remains a timeless, infinite, uniform *CONSTANT*, which may be ineffable as Newton describes (63).

In fact, compress Atoms and it's possible to produce a *DARK* Black Hole of concentrated Gravitatational Energy ($GE = MC^2$) *SPACE*.

Upon receiving a Nobel Prize, Max Planck nobly honored the "Matrix" of Source Energy that gravitates Atoms.

Life in **SPACE** is where Atoms, like us, physically bend, warp, or causally **EFFECT** our surroundings, which includes other Quarks and Molecules that affect us somewhat too.

The evidence is self-evident: Imagine one of us in a body of water (e.g. a tub, pool, lake, or sea of hemoglobin) where H_2O curves around △○□ different shapes by our presence alone. In fact, the larger we are – in physical size or non-physical (**informative**) scale – the greater distortion or gravitational pulling power, as is the case with celestial bodies or mass media.

INFORMation, which may be physical or (in)visible as spoken word, has gravitational potential energy to warp **MASSES** or an individual with Butterfly Effects. L👀K back @ significant times in HUMAN history and s👁👁 how **DATA** alone possesses super-massive power with **GFORCES** animated by Atoms.

At a super-massive scale, we're all in the **SPACE TANK** (69) on the mother Earth Ship following (193) our Super Star, while fellow Atoms warp, bend, and curve AROUND + (A)ROUND.

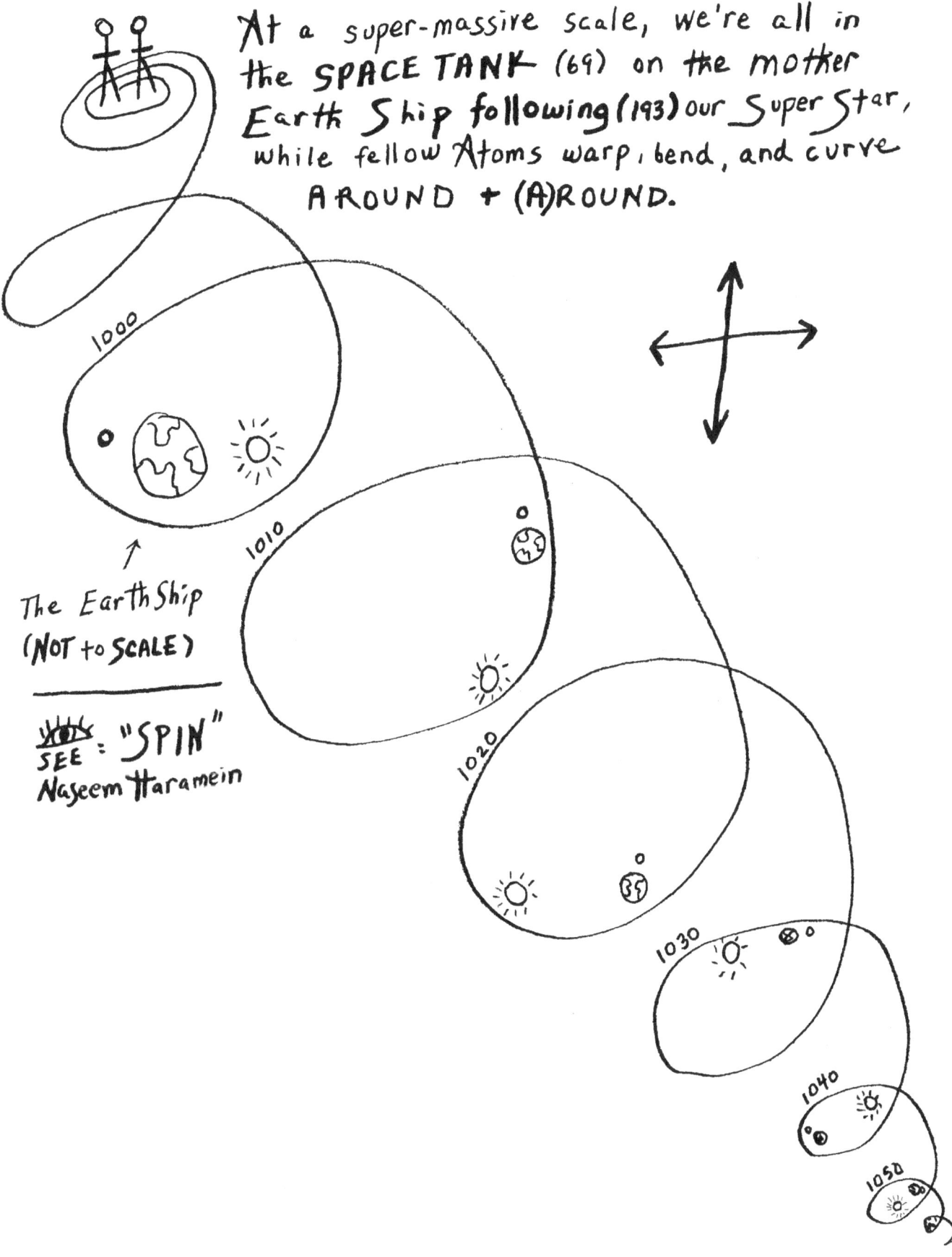

The Earth Ship
(NOT to SCALE)

SEE: "SPIN"
Naseem Haramein

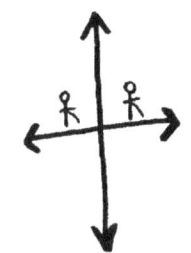

SPACE of Gravitational Energy is ISOTROPICALLY *INFINITE*, not "flat" like those examples shown of Atoms bending 'rubber sheets':

♥ Atoms BEND or warp by affecting other Atoms. Even our consumption of food, water, information or AIR—by breathing IN/OUT— gravitates fellow molecular structures.

On an episode of **INSECTIA**, Georges Brossard showed how different INSECTS gravitate like Réné Decartes' CLOCKWORK during the de(compost)ing process.

All of us — Fruitflies, Dust Mites, Bacterum, Molds, et al gravitate our **HOSTS** - but wildlife operates within highly advanced cashless societies.

As we know from, "The Secret" Law of Attraction, People attract— **or** repel each other, including material things and **ISSUES** as a causal effect.

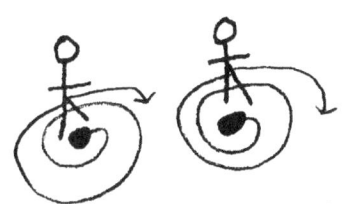

Our Gravity Well [as Melinda Gebbie described on the interview in "Mindscape of Alan Moore"] determines what flows into our universe. Regardless of who or what we are in a multiverse (141) of many, only that which we **MATCH** may attract.

Whether Butterfly, Bee, or the clever **Moth** that **MATCHES** Bumble vibration to access hive honey, All Atoms are governed by the same force of **SPACE** where we reside in numerous chemical forms. So, let it be known that STARS, planets, people, flora, and fauna are **NOT** living on a rubber sheet **RANDOMLY** gravitating each other.

As we know from "Strange Attractors" and "Chaos Theory" itself, anything may **SEEM** random without **INFORMATION** about the true **ORDER**.

OPTICKS: BOOK III — Isaac Newton

"It seems to me that particles have, not only a VIS inertiae accompanied with such passive **laws** of Motion that naturally result from that force, but also that they are moved by certain **Active Principles**, such as **Gravity** causing fermentation and cohesion of bodies.

These principles I consider, not as occult qualities, but as general laws of nature by which things themselves are formed; their TRUTH appearing to us by phenomena, not yet to be discovered.

To tell us that every species of things is endowed with an occult specific quality by which it acts and produces manifest effects, is to tell us NOTHING! To derive two or three general principles of **MOTION** from phenomena, and afterwards to tell us how the properties and actions of all corporeal things follow from those MANIFEST PRINCIPLES, would be a very great step in Natural Philosophy (science).

Now, all material things seem to have been composed of the particles mentioned by the counsel of an intelligent agent, who created and set them in **ORDER**. And, if so, it's UNPHILISOPHICAL to seek for any other origin of the world, or to PRETEND that it might arise out of a CHAOS. For while comets move in very eccentric orbs in **ALL MANNER** of POSITIONS, BLIND FATE could NEVER make all the planets move.

Nothing else than the wisdom and skill of a powerful, ever-living agent, who being in **ALL PLACES**, is more able to MOVE bodies within its boundless uniform sensorium, and thereby to form and reform the parts of the universe, we are by our will able to

MOVE the parts of our own bodies, And yet, are we not to consider the world as the BODY of **G**, and the several parts there of ?

A uniform being (**SPACE**), void of organs, members or parts ... with no need for such... being EVERYWHERE present to ALL! *IN FINITUM*, Unlike matter that is not necessarily in all places; it may also be allowed that **G** is able to create particles of several sizes and figures, in several PROPORTIONS to **SPACE** itself; and thereby ... making worlds of several sorts in several parts of the UNIverse."

Atomic structures are the ANIMATORS that bring all (in)visible things to life. Whether it's our words that directs behavior from INFORMATION, or our physical attributes that EFFECT others, Atoms bend, warp, curve, and (trans)form all MATTER (S) within the Gravitational *E*nergy of SPACE .

6

G WHIZ

Mergers + Acquisitions
SEEing is GEEing
Canticle to the Multiverses
US
The Signs

135

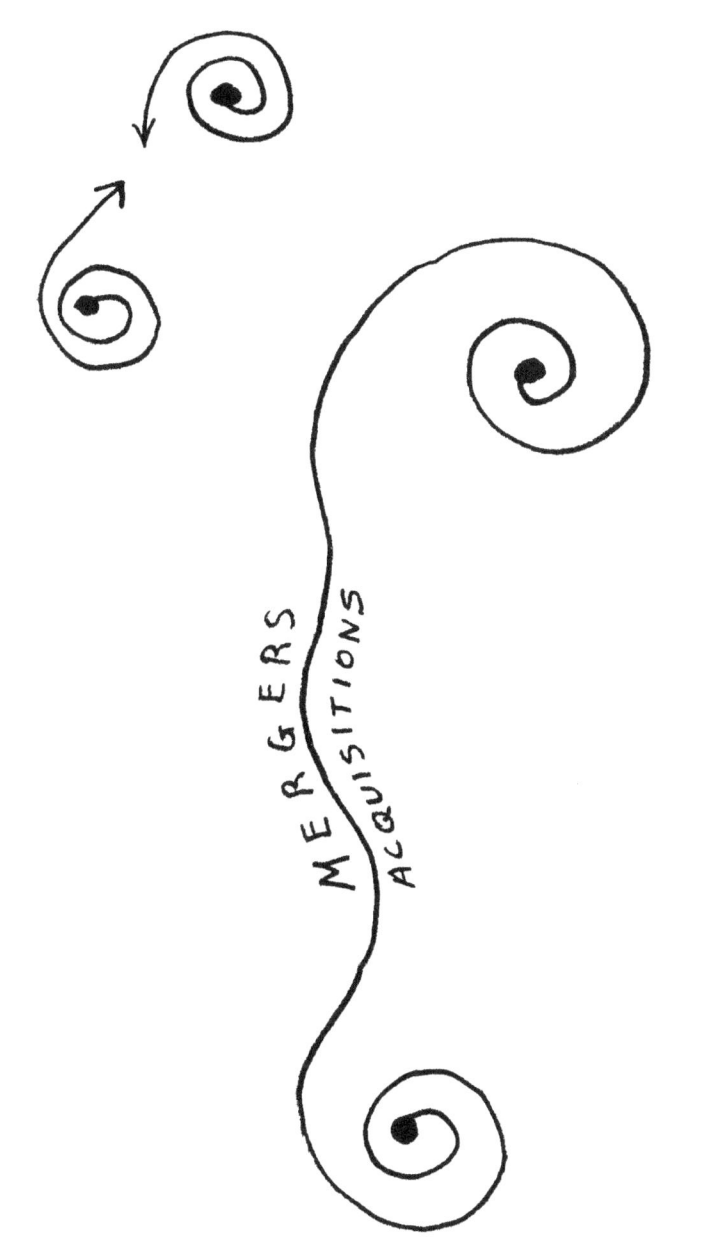

MERGERS
ACQUISITIONS

Celestial and Terrestrial Unifications
are self-similar at every scale.

@Joms™

Can Gravitational Energy (GE) be seen? Sure it's all around and within us, but maybe SPACE can only be scene when we close our eyes; although the 'LAMBSHIFT' evidenced GE.

I **SEE**, in the absence of Atoms, such as Star light, SPACE is the DARK (Energy, Matter, Flow) that we represent in physical form.

Regardless of scale, Atoms are brought together with the same GE source.

In fact, our daily nutrients look more like this:

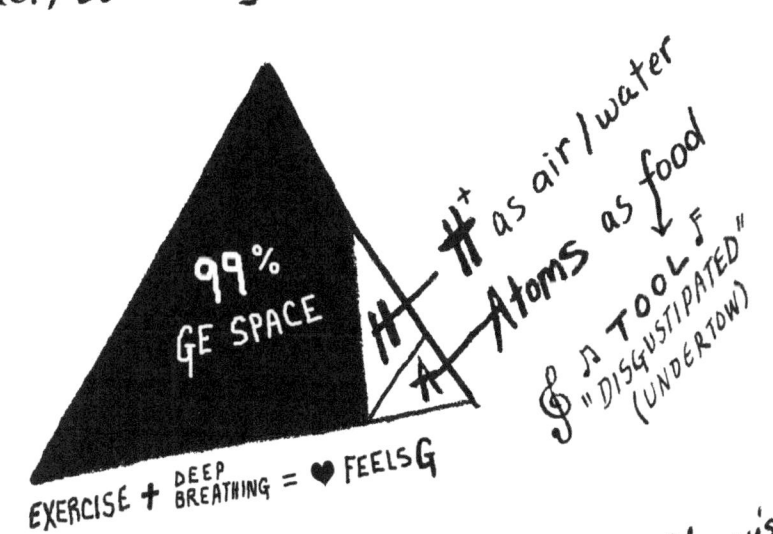

SPACE is the title of T. Henry Moray's book, "The Sea of Energy in Which the Earth Floats."

Indeed, Seeing is GEEing...

01

TETRAKTYS

@oms™

• CANTICLE TO THE •
MULTIVERSES

"Dragonflies are Identified Flying Objects (Ifos)."
—Joie

(Re) Member when Brian Swimme lectured about MULTIPLE UNIVERSES?

Oh Yea...
He mentioned parallel worlds of horses, haWKS, and insects.

We reside in the same area, as do humans with each other, but our multiverse is another dimension with neighbors on different WaYe lengths.

Can U PICTURE what the ATOMVERSE must be like?

? ? ? ?

DENOTES ∞
= zillions" of Atoms in SPACE(S)

•US•

"Data aequatione quotcunque fluentes quantites involvente
FLUXIONES invenire; et vice versa." —Isaac Newton

G Whiz

FLUXIONS represent everything Atom; it's the rate of our flow, change, and **MOTION** as calculated by Sir Isaac's Differential Equations; and this is why he famously proclaimed the importance of calculus **MOVEMENT**. In contrast to the 'constant' everpresent **SPACE**, we **MOVE** as the **MOTION** of the Universe.

Galilean Relativity also shows that Atom Movement operates in direct relation to other Atoms —since we're all scions of **H** — out of "one" with an infinite number of us connected with Hydrogen. Yet, from an Inertial Reference Frame, Gravitational Kinetic Energy gives us attraction - direction, along with **MOTION**, which is how we gravitate another.

$$R_{\mu\nu} - \tfrac{1}{2} R g_{\mu\nu} = 8 \pi G T_{\mu\nu}$$

The Principle of Equivalence in Einstein's General Relativity Theory (equation above), shows that Gravity — or Big **G** - is the same as **ACCELERATION**.

Since G is equal to ACCELERATION, the so-called Dark Energy, which is GRAVITY in disguise, continues to "ACCELERATE" expansion into itself.

Atoms are "ones" evolving and expanding, not the container of SPACE that remains constant. In fact, like celestial bodies, our change of velocity with time is often über-fast with invisible acceleration. Moving within the planet SPACE TANK or flying in an airplane at hundreds of kilos or miles per hour feels ordinary; acceleration is not obvious even when G'ing at high speed within a system.

As SPACE itself, G is the silent SHUNYA; a void where Atoms MOVE and beat our hearts on/off.

$$\heartsuit\ G^{IN} \qquad \blacklozenge\ G^{OUT}$$

Being dark, black, and invisible makes G force visible indirectly - thru Atoms. However, G's appearance as BLACK HOLES in its SPACE absorbs light completely into the black darkness of itself. Any Atomic Structure has the Gravitational Potential Energy (Abraham-Hicks' "vibrational Escrow") to form a BLACK HOLE if compressed below the Karl Schwarzchild radius:

$$r_s = 2\,Gm\,/\,c^2$$

US...

Made in its likeness, energy, and SPACE image,
G is US and we are the magic dividends of its
ZERO Point INFINTY :

$$\text{Atoms} \div \underset{\text{(zero)}}{GE} = \infty$$

Compounded Atomic Structures come in many molecules,
flavors, colors, sounds, shapes, and sizes with
numerous names given to US as we expand with new
"ones" born, mixed, transformed, invented or discovered.

Interestingly, whenever Atoms are divided by the
undefined singularity of GE, there's a potential
for an infinite number of US to compound in any form,
quantity or variety:

• Insects – How many kinds and subtypes are there?
• Flora and Fauna in general exists in great diversity.
• Other NOUNS, like Candy or Chocolates; Beers, Wines, +
  Beverages; or food and recipes are as endless
  as SONICS – songs of music – produced between the
  Silent SPACE of GE:

$$\text{♪} \div \underset{\text{SILENCE}}{\emptyset} = \underset{\text{INFINITE ATOMIC SOUNDS}}{\infty}$$

Through any of US – from the largest celestial mass –
to the tiniest terrestrial matter(s), infinite forms
are divisible by zero GE.

@Doms™
• THE SIGNS •

What do you think about ASTROLO**GE**E?

Some can (not) s👁👁 the self-similarities between celestial and terrestrial Atoms.

Looking ⬆️p, we see ♄'s clouds gravitating to and fro, in DIRECT response to "requests" in the Gravitational Cycle (43).

Wet reflections in the sky are mirrored on bodies of water during transit in the **SPACETANK** (69); and the same oceans + aquatic life are governed by ♄'s celestial tides (179) thru sun shine of day or sunny night.

★ There is, in fact, "ORDER OUT OF CHAOS," as Ilya Prigogine's book is titled.

Any phenomena or system appears random, without order, when **INFORM**ation is absent @ the observer's viewpoint.

7

MIND OF G

ENERGEE

SPELLS

ROM / RAM

B/W HOLES

$GE = MC^2$

@oms™
·ENERGEE·

Energy is GRAVITY; DARK invisible SPACE of which Atoms are 99% composed.

As the Mind of G, Gravitational Energy takes on many forms with an equal number of chemical names. However, regardless of the Atomic Structure, all has potential and Kinetic ability to move or WORK thru numerous capacities. Yet none of us are independant islands. Rather, we all gravitate one another thru LIKE vibrations @ different scales in the "UNI"verse. In fact, the word, COSMOS is Greek for ORDER, even though the SUMs of our world appears Random.

As a case in point, Atoms employed to produce the Butterfly Effect of WEATHER or climate as (H) sun, water, or wind, causally gravitates around Earth.

Want Acid Rain?

The recipe calls for certain Atoms in the air, produced primarily through human activity, perhaps. Mix Atoms well in the troposphere of our SPACE TANK (69). Add a dash of dust, sand, salt, or other condensation nuclei, and yoilà = Gravitation is done!

The SPACE of **GE** allows all...whether unwanted or desired. Hence, Abraham's book:

"Ask and It Is Given."

The only way to learn more about Gravitational Energy is to seek or study the left side of that famous equation, rather than focusing solely on the right "visible" side of $GE = MC^2$ (167).

@Joms™ GE SPELLS

- Gravity
- Energy
- Zero
- Invisible
- Isotropic
- Gravitation
- Kinetic
- Potential
- Work
- NO
- Yes
- One
- **SPACE**
- Ether
- Dark
- God
- Vacuum
- Vortex
- Black
- Open
- Quantum
- FlUX
- Life
- Atoms
- ALL

```
N  N  O  E  L  B  I  S  I  V  N  I
O  K  M  T  C  S  L  P  S  L  C  E
I  Y  U  H  I  E  L  A  M  A  I  F
T  G  T  E  T  Y  A  C  O  I  P  I
A  R  N  R  E  K  Y  E  T  T  O  L
T  E  A  L  N  T  L  E  A  N  R  M
I  N  U  K  I  K  R  A  D  E  T  U
V  E  Q  V  K  R  O  W  I  T  O  U
A  F  A  K  C  Y  S  R  Y  O  S  C
R  R  Z  E  A  D  Z  K  P  P  I  A
G  E  Q  N  L  O  X  E  T  R  O  V
Z  E  R  O  B  G  N  X  X  U  L  F
```

What do the words WRITE + SPELL have in common?

R ^W...

R $^{W}_{I}$ T
E

S $_P$ L
E L

? ♀ ?

♀ The common G's are INFORMATION = LETTERS
(common denominators)

If the English Alphabet was unknown, the data would remain scrambled like the riddle of 'random' Prime Numbers or any other system that appears CHAOTIC without information.

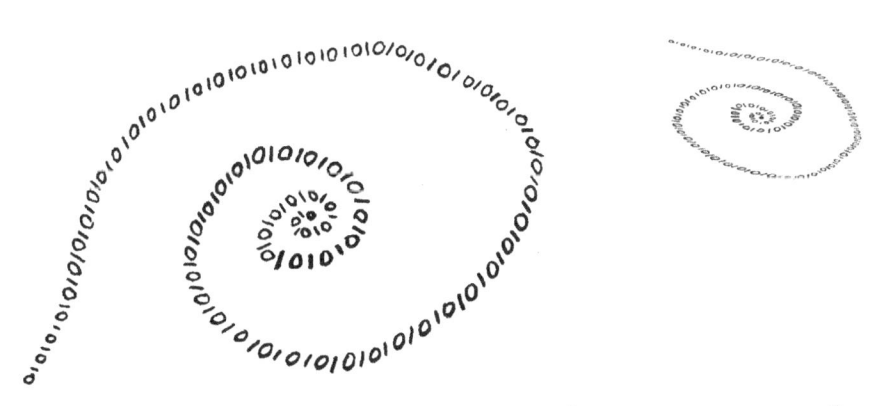

Universal Gravitation Spells Attraction of all Atomic Structures whom together unite and eventually Effect or transform each other in SPACE.

RAM is self-similar to Frontal Cortex Memory that represents dominate or active thoughts, which factors into how we **FEEL** (e.g. our VIBE).

ROM is more like storage of memories... info that may not be presently activated.

People process **INFORMATION** as CPUˢ in computers, yet we have sensory systems included with our divinely biological machines. In fact, the "Dog Whisperer," Cesar Milan often discusses how canines interpret our **ENERGY**. Although some humans are unaware of GE vibration, and how it determines interaction with others, we tend to gravitate different facets of the same nouns.

In Volume I, which was written mainly for consciencious LOA (79) practitioners, **JOIE** included a **GE** chart that reflects **POTENTIAL** gravitators based on cerebral activity. Often, when people are asked how they were **FEEL**ing 'around' the time of attraction, evidence shows frequency correlates with **GRAVITATIONS.**

"COMMON"
SUBJECTS GRAVITATE

♈ ◊ △ $ ⊕ ☉ ♫ = **INFO**

N.B. Offspring often gravitate PARENTS in relationships

("INFO" = Gravitation)

Isn't it interesting how most FORMS of InFORMation is dark, black, invisible, and unseen?

Data is all around us, yet much is hidden, and only appears or interacts with Gravitationally 'MATCHED' Atoms.

Bruce Lipton's, "BIOLOGY of BELIEF" directly relates to INFORMATION.
Yet, DATA are MAGIC SpELLs that possess 'POTENTIAL' GE to transform consumers, often instantly, post consumption.

What's even stranger is how spoken information is BLACK (invisible); and when it's transmitted via sound waves, past an event horizon, through vibratory senses of listeners; the data swirls, perhaps close to lightspeed, inside of our BLACK HOLES.

BLACK → HOLES

• Invisible Data $\frac{IN}{OUT}$
• No information LOST
• Light cannot escape

WoW! Do bodily expressions resemble WHITE HOLES where data or matter(s) exit after processing?

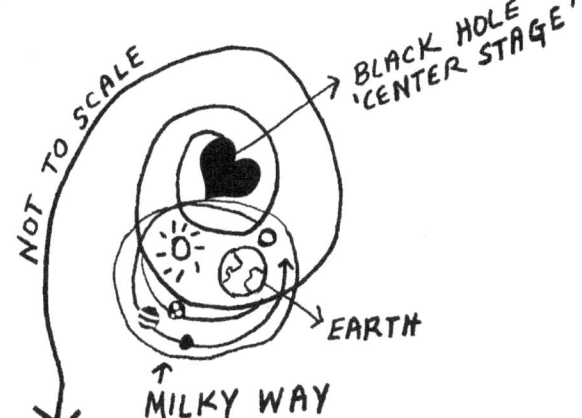

NOT TO SCALE

BLACK HOLE 'CENTER STAGE'

EARTH

↑ MILKY WAY

A **BLACK HOLE** concentrate churns with magnificent **GE** at the Milky Way's heart, where FOLLOWEES (193) trail the beat resonating from the center spin.

Zoom in to an observer as a BODY in **SPACE** and their local superpositions are not obvious. Hence, our Atoms are in multiple places, interacting simultaneously, in unison as a w**HOLE**.

Fred Alan Wolf in a "Down the Rabbit Hole" animation referenced us as residing in a PAC-MAN game, where most rarely see outside of the simulation.

Since we're all **part** of the wHOLE, calculations often show that **part**icles can be in more than the 13 places listed on the right; and it's evident by SCALE. Atoms can be in a body of water, on a farm, within a town, country, continent. and other **SPACES** at the same time.

Everything we do as entangled COLORS (39), ripple BUTTERFLY EFFECTS, and writes DATA to the SPACEBANK (183) of INFORMATION, which remains EMBEDDED in the cosmos throughout Atomic (trans) formation.

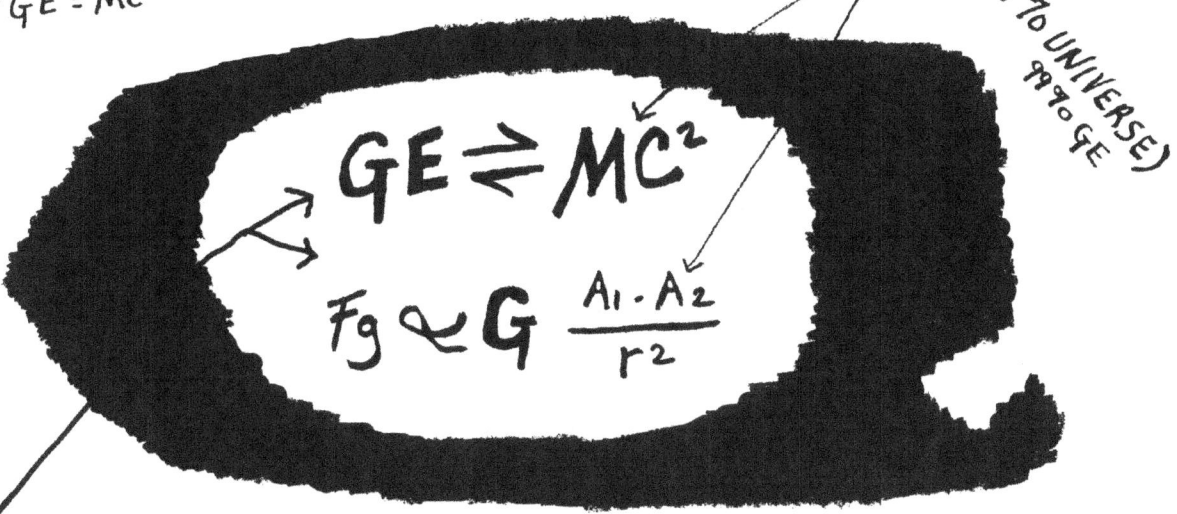

@)oms™
•GE=MC²•

Atoms (4% UNIVERSE)
99% GE
CONSCIOUS

$$GE \rightleftharpoons MC^2$$

$$F_g \backsim G \frac{A_1 \cdot A_2}{r^2}$$

Gravitational Energy:

0 SPACE: ALL PLACES + IN ATOMS

1 ENERGY: ALL FORMS $\rightleftharpoons$ ZPGES (97) CONSCIOUSNESS

2 LAW OF ATTRACTION : GRAVITATION (79)

3 ACCELERATION + INERTIA : CONTROLS ATOMIC MOTION

4 INFORMation : ALL OF THE ABOVE + SPACEBANK (181)

5 OTHER NAMES : TIDAL FORCES, CORIOLIS EFFECT, ETC (15)

U N I F I C A T I O N

Feel that ?
SPACE is in us all. Gravitational **Energy** resonates
(in) Atomic Waves; it's **GE**motion that decendants of
Hydrogen (53) 'SIGNAL' as we gravitate each other
@ EVERY scale, including the SPOOKY Quantum level.

8

OH G

PISCESOREAN THEORUMS

GRAVITATIONAL LENS

SPACE BANK

PT's I-III

Piscesorean

Theorums

$\mathcal{L}$ $\mathcal{K}$

NEIN - EINSTEINIAN

 @Joms™

• P.T. I : ISSUES •

PUPPETS FROM SPACE

We watched only the first ≈ 2☺ minutes of this animation; in it, **FISH** ⚯ α were the wise ones, which leads us to the concept of **PISCESOREAN THEORUMS.**

Our first nein-Einsteinian one is :

$$PT\,I = \frac{FISH}{F \big| ISSUES}$$

FISH TISSUES, like Avian Issues, tell us how we causally EFFECT them in the SPACETANK—— (69).

These Orcas appear to be swimming in opposite directions, but are actually gravitating together, as fractals.

We're all performing self-similar motions on a GRAND SCALE of **CONTRASTS**.

EVERY EVENT ON Earth has a beginning (29) or starting point; it's the YIN / YANG of our colorful (39) CONTRASTS that generates change and desire, which yields improvement or EVOLUTION (195) VIS VITA. ♺

W
EAST
S
T

LI ☯ QI

F
MALE
M
A
L
E

PT. III : THETA MYSTERY SOLVED?

THETA

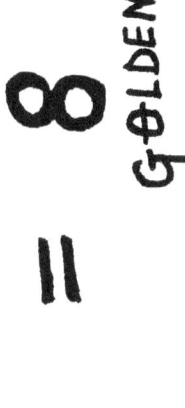

SPACE

BLACK

ATOMS

RED
ORANGE
YELLOW
GREEN
BLUE
INDIGO
VIOLET

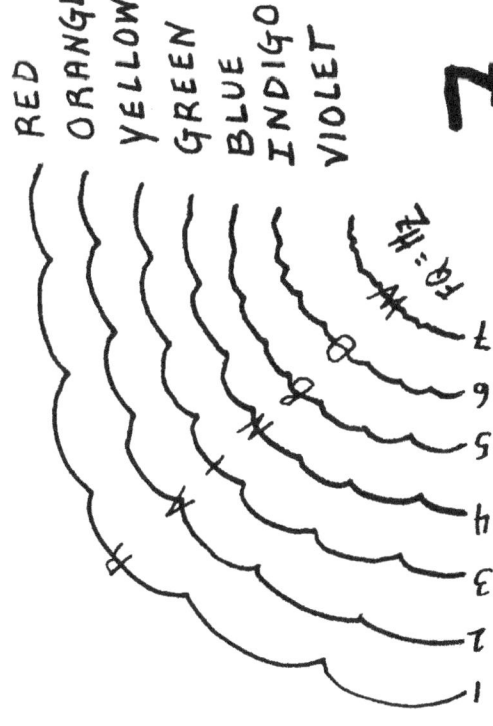

1 2 3 4 5 6 7

FOR = Hz

7
VISIBLE "LIGHT"

+

1
INVISIBLE "DARK"

= 8

GOLDEN SPIRAL

GE SPACE LOOKS LIKE THE PLACE AND SACRED MEDIUM FOR **SPIRALITY**, AS INDICATED BY DIVISION OF 233/144 IN THE FIBONACCI SEQUENCE, WHICH PRODUCES THE GOLDEN RATIO=1.61803399 SPIRAL. HENCE, OUT OF 1, MANY FORMS, COLORS, POLYTHEISMS, AND FRACTALS... AD INFINITUM.

@oms™

Worn by the Moon!

GRAVITATIONAL LENSES
★ THE LENS THAT GRAVITATES ATOMS ★

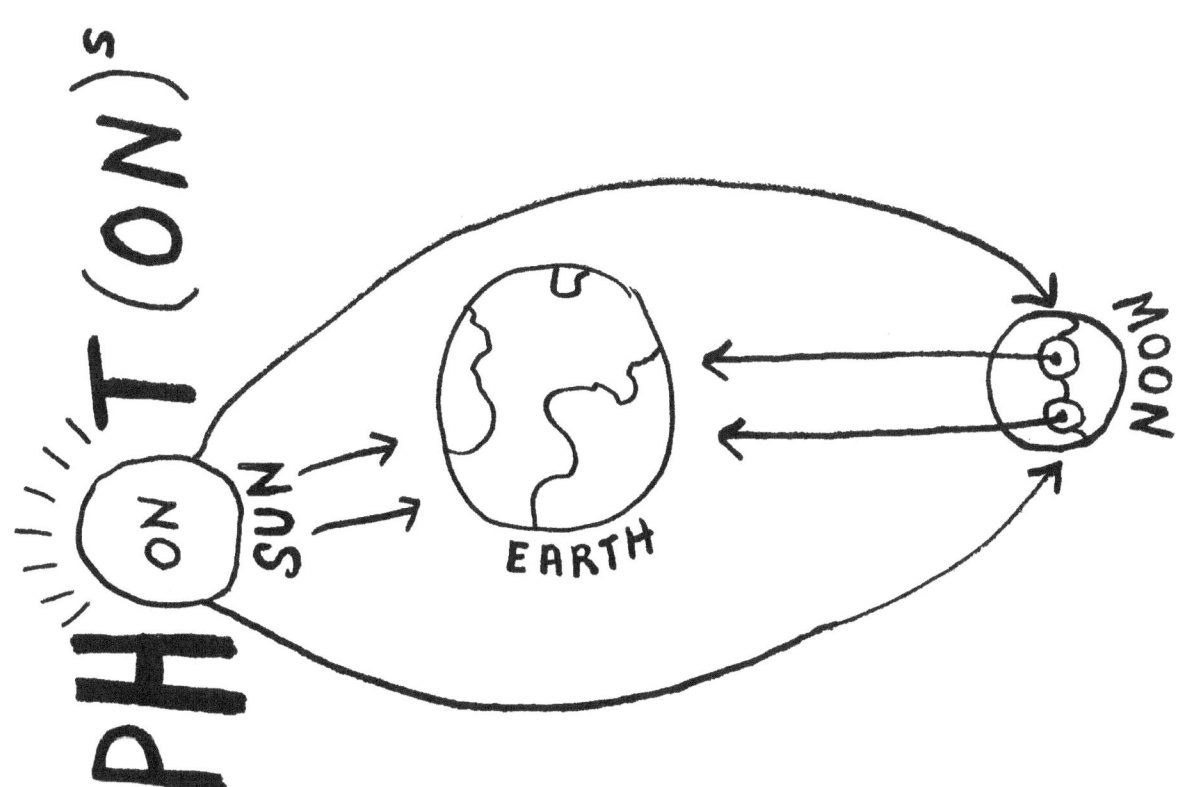

PH T(ON)ˢ

SUN · ON

EARTH

MOON

Where **GE** signals, ⇅'s *FLUIDS*
and PHOTONS doth **GO** ...
to places and **SPACES**
that Atoms do flow.

@oms™

SPACE BANK™

THE OFFICIAL INFORMATION CLOUD

It's Wireless!

INFORMATION exists all around us, including answers to questions and dilemmas, like the Riemann Hypothesis; data are electrically floating in the aether of **SPACE.**

As we know, everything that physically exists around us is HISTORICAL, since it manifested in the past, and holds MEMORY (ROM + RAM (161)) through individuals, owners, or observers. Yet, because the physical world seems so real, many do not tap into the invisible World Wide Web.

← The web or fabric of SPACE is embedded with info.

← As a case in point, the **BOLD** represents data a person "KNOWS" (e.g. other people, resources, etc).

SPACE Knows <u>ALL</u> since everyone and everything— that has ever been or is to be — is contained within its body. The multidimensional **ZPGES** (97) is Somewhat self-similar to a SUPER GOOGLE, but it's the best "Atomic" Search Engine that's able to connect users @ Spooky distances across the web of life. Plus, as referenced in, "The Secret Gravitational System" by Joie, **GE** signals Atoms with impulses that Newton duly describes (15), which is also Known as instinct, gut feelings, intuition, or what others term, "COINCIDENCE" that ironically means SYNCHRONICITY.

Others have concepts surrounding Information Theory, such as the *Holographic Principle*, IDEA SPACE, and another is the "AKASH" or Akashic Record. The **SPACE BANK** is our new term that describes the invisible "wireless" cloud of **INFORMATION** existing everywhere; it refers to data about everyone and everything in existance. People, places, and things are absent of inherent meaning; viewers assign data to nouns, including themselves, although the quantity of INFO is INFINITE with unlimited allocative resources.

The other term we invented is **SPACE BYTES**™ since there's an endless stream of Atomic info simultaneously processing via innumerable Sources. **Most data are DARK** as energy (GE); it's invisible as feelings or instinctive signals that offer direct guidance and gravitational synchronicities.

INFORMATION RULES, and we are the processors with personal **SPACEBANK** accounts in the Mind of G that's self similar to the operation of LIBRARIES. In **SPACE**, data are instantly accessible [to add to the collective body of KNOWLEDGE] through our deposits, and withdrawls of information.

As an anesthesiologist, Stuart Hameroff discusses the "UNCONSCIOUS" and how LOCAL or personal INFO remains in the universe even if patients transition.

⊕ **SPACE BANK ACCOUNT # 010.000.110.010.111 ...**

PERSONAL RECORDS
(WOVEN IN SPACE)

011011011010110,
01110111010101110101001
111011100011110110001100
1010101011011011001100
11000001100010010100

PATIENT
(Physical BODY)

Gravitational Energy beings are scions of **H** (53), and everything we do remains as 'parts' of the cosmic mix ... continuously (trans)forming and evolving **EVOLUTION**.

The Quality of Life or amount of information users engage depends on who and where they are, which is a byproduct of their requests and desires (83). For instance, a BIRD BANK ACCOUNT is based on that species, which determines where they go, what they do, see, or consume, and with whom they interact. Fowl, like other wildlife follow direct "GE" INstinct that leads them to their relative foodchains, migrations, or hibernation, if applicable, and so on.

PEOPLE are far more complex, with supercomputer brains that generate numerous querries and other data that's also stored via cells and sensory systems.

• SPACE BANK...

PERSONAL INFORMATION may include, but is naturally not limited to, stories told or received, books read, songs, sounds, sights, and other inherited knowledge or experience. RECORDS we "**HAVE**" often determines what is engaged, unless new information is streamed outside of our ACCOUNT.

In recent times, evidence for the usefulness of meditation has been shown, although utilized for centuries. It is known that **SILENCING** auto-thoughts results in divine **SYNCHRONIZATION** with GE; the 99% and 1% = 100%, figuratively speaking. Psychedelics, Hypnosis treatment, and other alternative **STATES** connect to multiple dimensions of **INFORMATION** inherent in **SPACE**.

Yet, MEDITATION is easy, free, and instant with ZERO side effects (pun intended); it's almost self-similar to writing ZEROS to a computer drive in some cases.

⊖ SLEEP mode does the same at a deeper zen level where one's internal GUI produces feature films
0-OFF that we call dreams (51).

😮 Conversely, when most humans are ON, it's like the view of city lights from **SPACE**; data thru active thoughts
1-ON and other MEDIA streams non-stop. →↑↓←

SPACE is a magnificent place (209), as are we, living as its body parts.

"PARABOL (A)"

TOOL

LATERALUS

9

G BLESS

FORM
FOLLOWEE~~E~~
EVOLUTIONISM
ATOMISM

@Joms™
• FORM •

Wow, l👀k @ that FORMATION!

Maybe it's a FOWL SUPERCLUSTER?

ATOMS FRACTAL "UNIVERSALLY" GRAVITATIONALLY

• FOLLO**WEE**

Follow We?

FOLLOWER   FOLLOWEE

The new science of **GRAVITOLOGY** (201) explores fractal attractors. In addition to resonating self-similar Gravitational Energies (e.g. Emotional States) or Electron Configurations, what else do **FOLLOWERS** and **FOLLOWEES** have in common?

FOLLOW THE **FOLLOWING** FOLLOWERS: ↩

• MOON   FOLLOWS   EARTH
• EARTH   FOLLOWS   PLANETS + SUN
• SUN   FOLLOWS   STARS + Milky Way (MW)
• MW   FOLLOWS   GALAXIES + ANDROMEDA (A)
• A   FOLLOWS   LOCAL GROUP (LG) + (MW)
• LG   FOLLOWS   • • •

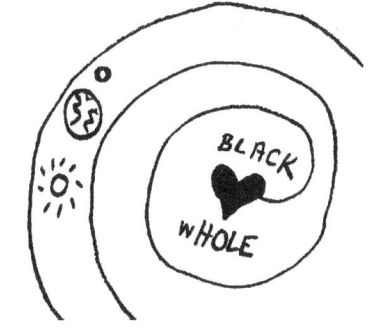

BLACK WHOLE

•EVOLUTIONISM•

IS the universe inflating or are some Atoms simply moving away — in contrast to those (like Andromeda) whom are cosmically deflating into the Milky Way ?

Hmm...

Maybe INFORMATION (181) is the only substance EXPANDING since no one knows if the BIG BANG Hypothesis is true?

Well, we do know that gravitational attraction + repulsion is a natural ebb + flow of contraction and expansion in evolutionism.

→ ← ← →

@Soms™
•ATOMISM•

♀ Hey Atom!
Yeah Atom——? ♂

Ask any Atom a question, and the answer is usually given in atomic language.

There's Atomic Time — based on the Atomic Clock or Atomic Calendar, and there's even Atomic Numbers in the Atomic Age.

Ⓐ
Ⓣ Ⓞ
Ⓜ

In other words, we (can)not see ▮▮▮ or speak outside ourselves, which may be why our own SPACE remains hidden - invisible to Atomists — even at the Atomium.

When Atoms are ready, data appear.
When Info is ready, Atoms appear.
When Observers are ready, Gravitational **SPACE** appears.

10

COLOPHON

GRAVITOLOGY

(K) NEWS

12. 12. 12

"*TRUTH* is the offspring of **SILENCE** and unbroken *MEDITATION.*"
— Isaac Newton

GRAVITOLOGY explores **ENERGY** (153) as **GRAVITY** in numerous **SILENT** forms (97, 167). In fact, Atomic Structures (molecular + sub**atom**ic) possess the same 99% **GRAVITATIONAL ENERGY** (GE) **SPACE**.

Humans are **HOLLOW** with "*The Secret* (79)" non-physical **GE** infiltrating all of our anatomy; that includes 'EMPTY' BONES and wet-waterproofed (53), flexible machinery. The primary engineering that sets us apart from other species is our supercomputer brain, which chooses (117) processes, and evolves **INFORM**ATION. Yet, as members of an intelligent system [that those like Bernard Haisch define as "God Theory"], **SPACE** is the medium where we all come together in the Gravitational Cycle (42).

Interestingly, like Atoms, CELLS also gravitate thru use of **electron**ic signals. THIRST, for example, is a 'CALL' that we 'ANSWER' with **H**ydration. Although it seems THE DRINKER desired H_2O alone, there's an intricate hierarchy of organisms (91) co-creating in **SPACE**s of the w**HOLE** (165).

In fact, **H**ydrogen itself is INFINITELY FRACTALLED as our most abundant ancestor, residing at virtually EVERY SCALE of the universe with multiple **electron** configurations, which may be the reason we're so connected.

By examining Gravity (BIG "**G**") as the energetic **SPACE** that contains "us" physical parts, we may finally gravitate UNIFICATION.

♥ @² HRH

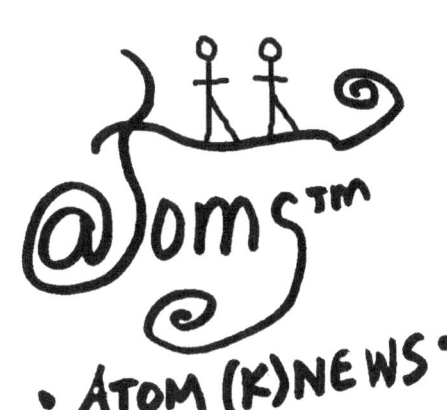

• ATOM (K)NEWS •

...THE ONLY KNEWS
BY ATOMS
FOR ATOMS

Shh!
It's starting...

NOW... 205

ATOM ᴋNEWS™

♪ Nouns Sing ♪
Atomania ♫
𝄞
@toms™

AP: VICTORIA—ATOMS, HRH CELEBRATE A NOBLE VICTORY AFTER ANNOUNCING FINDINGS ON THE, "THEORY OF EVERYTHING ATOM:"

$$GE \rightleftharpoons MC^2$$

Atomic Press

IN AN EXCLUSIVE INTERVIEW, MOL^S ASKED ABOUT THEIR FAIT ACCOMPLI, AND THE @toms SAID, "IF **FISH** FEEL, PISCES KNOW," WHATEVER THAT MEANS?

WE'LL HEAR MORE FROM THEM AFTER THEIR WORLD WIDE TOUR AROUND THE WEB.

BON VIVANT!   BON VIVEUR!

ANNUAL
SPACE DAY

12.12.12

12:12 − 12:12

ANNUAL
SPACE DAY

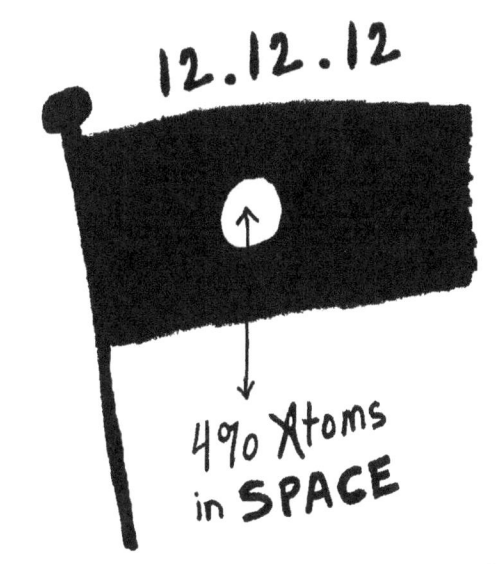

12.12.12

4% Atoms
in SPACE

🕐 SPECIAL REPORT:

Since 2012, SPACEDAY is celebrated on 12.12 PLANETWIDE from 12:12 on one side of the Earth to 12:12 on the other, which commissurates Atomic unification with SPACE.

"BOOM!" S.O.A.D.

"From Gods to Men
From Atoms to Worlds

From *Stars* to Vital Heat
of Organic BEINGS

From Existance and Form
is an IMMENSE CHAIN
Whose LINKS are ALL

CONNECTED."

— Madame H. P. Blavatsky

BY Joie
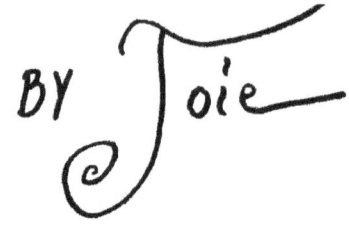

VOLUMES

I "The Secret" Gravitational System

II GRAVITY = Conversations with **G**
A Common Dialog on Universal
Gravitation As "The Secret"
Law of Attraction

★ FEATURING: THE @Joms™
A Graphic Novel of Pure Abstract
(NON) Fiction

III PRINCIPIA + OPTICKS
THE SECRET DOCTRINES

FILM **GRAVITOLOGY**

MUSIC HARMONIES OF THE WORLDS